**PERSPECTIVES IN COGNITIVE NEUROSCIENCE**

Stephen M. Kosslyn,
SERIES EDITOR

# NEURAL PLASTICITY

The Effects of Environment

on the Development of

the Cerebral Cortex

## Peter R. Huttenlocher

**HARVARD UNIVERSITY PRESS**
Cambridge, Massachusetts
London, England 2002

*Library of Congress Cataloging-in-Publication Data*

Huttenlocher, Peter R.
Neural plasticity : the effects of environment on the development of the cerebral cortex
/ Peter R. Huttenlocher.
p.   cm.—(Perspectives in cognitive neuroscience)
Includes bibliographical references and index.
ISBN 0-674-00743-3 (hardcover)
1. Developmental neurophysiology.   2. Neuroplasticity.
3. Cerebral cortex.   I. Title.   II. Series

QP356.25 .H88   2002
612.8'25—dc21                2002017106

# CONTENTS

## ACKNOWLEDGMENTS

This book would not have been possible without the unflagging support and encouragement of my wife, Janellen Huttenlocher. She critically read key sections of the book. Perhaps more important were the many stimulating discussions we had related to the subject matter of the book. These provided me with many helpful suggestions. My son Daniel helped me in my attempts at mastering word processing. I wish to thank Dr. Arun Dabholkar for contributing several of the illustrations. He also provided valuable technical support during the more recent phases of the synaptogenesis project. Angela Song was an expert editorial assistant, and she in particular was essential for getting the bibliography into shape. Marie Sheldon provided much-needed secretarial help.

Finally, I would like to remember the late Hendrik Van der Loos, who stimulated my interest in developmental neurobiology, and specifically in neural plasticity. A key portion of my work on synaptogenesis was done during a sabbatical stay at the Institute of Anatomy in Lausanne, Switzerland, where Hendrik was the director at the time. He was a gracious host and a fine teacher and mentor. I miss him very much.

# NEURAL PLASTICITY

# INTRODUCTION

The question whether and to what extent environmental input influences the development of the child's brain has generated much recent interest and discussion. The old "nature versus nurture" argument is being revived, and proponents on both sides are often making use of new tools from a number of related disciplines, including cognitive psychology, developmental neurobiology, brain imaging, and computer science. The subject has been addressed via animal models and computer simulations of neural networks as well as in studies on humans, both in normal and in developmentally abnormal or deprived populations. This work is widely dispersed in various disciplines and reported in specialty journals. The purpose of this book is to provide an overview of what is known concerning the effects of environmental input on the structure and function of the developing brain, and to develop as coherent a point of view on the subject as is possible at present.

Both positive and negative effects of environment on brain development have been described. At the level of cognitive functioning and behavioral adjustment, detrimental effects of environmental deprivation during childhood are being increasingly recognized. Several remedial and enrichment programs for such children have been critically evaluated (Ramey and Campbell, 1984, 1994; Campbell and Ramey, 1994, 1995). But questions as to what kind of environmental input would be most effective, or the age range over which the greatest sensitivity to input can be found, are not resolved. In particular, there is controversy over whether a time window exists during which remediation is more effective than at other times, as well as over the age limits of such a time window (Bruer, 1999).

A related issue is that of the timing and effectiveness of remediation programs for developmentally handicapped children, including children with cerebral palsy and with cognitive impairments. In the child with focal brain damage, the available evidence shows a remarkable ability of residual normal cerebral cortex to "take over" functions normally carried out by the damaged regions. However, this malleability of cortical representation also appears to have negative consequences, which have been ascribed to the crowding of multiple functions into a smaller than usual volume of normal brain tissue (the "crowding hypothesis," Hebb, 1949). This crowding effect may account for the observation that brain damage in infancy—focal as well as diffuse—tends to lead to impairment in the overall efficiency of cortical functions, reflected in a decrease in IQ. An infant with only a single normal cerebral hemisphere is able to carry out most of the basic cortical functions of both hemispheres, no matter which hemisphere was injured, but he or she performs these functions less well than a normal infant. At this point, we have very little information concerning the effects of remediation programs on neurologic and cognitive functions in children who incur brain damage in the prenatal, perinatal, and childhood periods.

A final question currently being debated is whether—and to what extent—enriched environments have an impact on the intelligence, school achievement, and success in adult life of children growing up in apparently normal homes. Should children be taught second and third languages? If so, at what ages? The same questions can be asked about other activities, such as intensive music training and physical training in, for example, gymnastics.

Some of these questions asked about early enrichment programs, including what kinds of programs are most effective, whether they carry over to adult performance, and even whether early stimulation programs may have negative effects, can be and are being addressed by carefully designed, longitudinal studies of children exposed to different curricula. Some of these issues have been explored in animal models. Studies in rodents have shown effects of various types of environmental enrichment on the synaptic organization and function of the cerebral cortex (Green, Greenough, and Schlumpf, 1983; Greenough, Black, and Wallace, 1987; Kleim et al., 1996).

An exciting aspect of the study of cortical plasticity is the increasing relevance of experimental developmental neurobiology to issues of human development. Specifically, developmental neurobiology is beginning to provide data related to the neural basis of environmental effects on cognitive function. These data in turn may have relevance for questions related to the existence and age limits of "critical periods," or "windows of opportunity" during which intervention is likely to be most effective, as well as for the types of intervention that are most likely to have positive effects.

This book provides a summary of relevant data from a number of fields of inquiry: developmental neuroanatomy (Chapters 1 and 2), computer science (Chapter 3), neurophysiology (Chapter 3), and brain imaging (Chapter 3). Chapters 4–6 provide more detailed discussions of the best-known examples of neural plasticity in each of three systems: sensory, motor, and language. Chapter 7 deals with plasticity in executive (reasoning and judgment) functions. Chapter 8 is concerned with plasticity in adult cerebral cortex, and compares the malleability of the adult brain to that of the infant. The two final chapters discuss correlations between the neuroscience and human developmental data, and emphasize what data from the neurosciences may tell us about human brain development and its modification by environmental factors.

A central hypothesis underlying remediation and enrichment programs is that the brain is more malleable during infancy and early childhood than later in life. This malleability leads to an increased capacity for learning, which in turn provides an opportunity for the improvement of cerebral functioning that cannot be reproduced to the same extent or with the same ease later in life. This property of the immature brain is referred to as neural plasticity. The neurobiologist Marcus Jacobson, in the third edition of his classic text *Developmental Neurobiology,* says that the term "plasticity" is used to refer to certain types of adjustments of the nervous system to changes in the "internal or external milieu" (1991, p. 199). Jacobson's definition of neural plasticity implies its occurrence under two conditions, namely (1) related to adjustments of the normal, developing nervous system to changes in the "external milieu"—changes in input through the sensory systems (visual, auditory, somatosensory, olfactory, and autonomic); and (2) related to adjustments of a nervous system to changes in the "internal milieu," such as those that would occur after focal brain injury. Plasticity related to focal brain injury would be secondary to the reaction of the remaining normal tissue to lost or altered input from damaged regions. It could either improve function or increase functional deficits beyond what would otherwise be expected. Examples of both types of plasticity are provided later in the book. Adjustments of the damaged brain will be referred to as "lesion-induced plasticity."

An important factor that needs to be stressed is the fact that malleability of the nervous system does not end with maturity. It has been shown to persist to some extent until old age. In order to prove the existence of increased plasticity in the immature brain one therefore has to compare the effects of a given intervention on the function of the immature brain to the effects that occur in the adult. Plasticity of the adult brain has been the subject of intensive recent studies (Julesz and Kovacs, 1995). This work is considered in Chapter 8.

Plasticity has been described in many neural systems, but it is most evident in immature cerebral cortex. A classic example of cortical plasticity related to altered sensory input is the rearrangement of geniculo-cortical connections that occurs in response to a change in visual input, due either to unilateral visual deprivation or to strabismus (Chapter 4).[1]

In the infant, unilateral visual deprivation leads to weakening of the synaptic connections between the deprived eye and the primary visual cortex (area 17) and strengthening of the connections from the eye exposed to normal visual input. A similar effect is not seen in the adult. The functional effects of strabismus (squint) that has onset in infancy are very different from those of dysconjugate gaze of adult onset. Strabismus in the infant leads to suppression of the image from the squinting eye, and in the untreated case results in permanent decrease in vision in the squinting eye. No such suppression of the image occurs in adult onset squint, which causes a rather disabling symptom: persistent diplopia (double vision).

Lesion-induced cortical plasticity may lead to the "taking over" by spared cortical areas of functions normally subserved by the injured tissue. This effect is most pronounced in response to unilateral brain lesions in early infancy. A striking example is the relative lack of language deficits in infants

---

1. The terms "strabismus" and "squint" are used interchangeably to describe a common eye movement abnormality in childhood in which the eyes are dysconjugate (for example, either converge or diverge). Each eye moves fully when tested alone. Squint is a developmental disorder of the visual and oculo-motor (eye movement) systems that mediate binocular interactions. Binocular vision involves the execution of finely coordinated movements of the two eyes such that a retinal image is always kept in the same location on the retina in both eyes. Children with strabismus fail to develop binocular vision and as a result have impaired depth perception (stereopsis).

with damage to the dominant (left) cerebral hemisphere (Chapter 6). For example, a child 3 years old with an acute left hemisphere lesion that destroys the major speech areas, and even a child in whom the entire dominant cerebral hemisphere is surgically removed, will be mute for a few days, but will then regain language very rapidly, almost to a normal level. An adult with a similar lesion will lose all or most of his or her speech comprehension and production and often has little or no return of language functions, even with intensive speech therapy.

Very recently, plasticity has also been demonstrated in normal language development. A prevailing view, that language, and especially grammar, is largely innate, is being challenged by data that show large effects of various environmental inputs on language development in infants and young children.

While neural plasticity probably exists in the nervous systems of all species, it appears to be most marked in specific regions of human cerebral cortex, in areas that subserve the so-called higher cortical functions, including language, mathematical ability, musical ability, and "executive functions." Regions of the cerebral cortex that subserve voluntary motor activity and primary sensory functions, such as visual and auditory information processing, appear to be less malleable. For example, unilateral damage to the left cerebral hemisphere during infancy will not result in major language difficulties but will produce right-sided weakness (hemiparesis) when the motor cortex is affected and a right visual field defect (homonymous hemianopsia) when there is involvement of the visual cortex. Yet some age-related plasticity is demonstrable even in these simpler systems. Examples include the absence of facial weakness and the presence of motor activity in the paretic arm in the form of mirror movements in children with perinatal unilateral motor cortex damage (Chapter 5), and strabismic amblyopia (as mentioned above) in children with squint or strabismus (Chapter 4). Plasticity in these simpler cortical functions has the advantage of being amenable to study in animal models, and such studies have been done beautifully for strabismus and to some extent for neonatal motor cortex lesions. These studies have made it possible to obtain vital information concerning the physiologic and anatomical rearrangements that underlie the observed effects on function. Animal models are not possible for the study of plasticity in the uniquely human higher cortical functions, except perhaps for the study of bird song, which provides a model for limited aspects of human speech (Chapter 6). In this field, more than in other areas of neurobiology, investigation of humans is critical.

The study of the effects of environmental input on the anatomy, physiology, and function of the nervous system is of relatively recent origin. Previously, genetic ("innate") factors were often cited as being the most critical for the emergence of higher cortical functional organization, including the emergence of the neural substrate for overall intelligence (Herrnstein and Murray, 1994) and speech (Pinker, 1994). Undoubtedly, genetic programs are of great importance. They predominate in the early steps of cortical development, including the birth of neuron precursors in the germinal matrix (periventricular) zone of the embryo, specification of these cells toward either neuronal or glial differentiation, neuronal migration from the germinal matrix to the developing cortical plate, the growth of dendrites and axons, and the formation of early synapses (Chapter 1). The discovery of the genes that regulate these developmental programs is an exciting area of developmental neurobiology. At the same time, however, it has become evident that genetic programs cannot account for all aspects of brain development.

The mechanisms by which environmental effects are exerted on the developing brain have been defined only recently, and much is still unknown. An exciting set of findings suggests that the environment has effects on the synaptic organization of the cerebral cortex, which underlies cortical function (Chapter 2). The types of connections or neuronal circuits that form appear to be to a significant extent determined by environmental input, especially during the postnatal period of development, which includes the early childhood years in the human.

Karl Lashley (1929, 1950, 1951), and Donald O. Hebb (1949) were early proponents of strong environmentalist views of brain development (Chapter 9). However, developmental neurobiology had not developed sufficiently in their day to provide a solid footing for their views. More recently, a theory as to how environmental input might affect the organization of neural systems was developed by the French neurobiologist J. P. Changeux (Changeux and Danchin, 1976; Changeux, Heidmann, and Patte, 1984). Changeux reasoned that the human genome is not large enough to provide for exact specification of each of the billions of synapses that are formed during brain development. Many of these contacts appear to occur at random. The question he posed was: How can organized function emerge from a nerve network with random connections? He suggested that this is possible through input to the system that has recurrent patterns. Such input, at first from sense organs to primary sensory areas, later from primary sensory areas to association cortex, was postulated to lead to functional specification of some of the randomly formed synaptic contacts. These synapses, which become incorporated into functioning neuronal circuits, become stabilized and persist. Synaptic contacts that are not utilized in neural circuits disappear.

Changeux's postulate gets considerable support from recent data: First, it has been found that electrical activity is indeed important for the persistence of synaptic contacts. Absence of activity leads to weakening of synaptic connections. Second, the scheme of synaptogenesis proposed by Changeux requires the random production of large numbers of synapses, some of which will become stabilized while others will disappear. This implies that at the time of the formation of synaptic circuits, the total number of synaptic contacts should exceed the number that is present in the mature state. A period in development during which the number of synapses is increased above that in the adult has been found in many systems, and is a prominent aspect of cerebral cortex development in humans and in other primates (Chapter 2). Third, if sensory input is important for the formation of neural circuits, it should be possible to influence the type of connections that are made by modification of the input to the system from the external environment. Evidence for such effects exists, primarily in the visual system (Chapter 4). Fourth, finally, it has been possible to reproduce the development of organized activity in a computer simulation of a random nerve net that is exposed to repetitive input. This is beautifully demonstrated in Gerald Edelman's 1987 book *Neural Darwinism* (fig. 7.5, p. 190), and in Jeffrey Elman's work on "language learning" by a computer simulation of a neural net (Elman, 1993; Elman and Zipser, 1988; Elman, Bates, et al., 1996).

The last chapter of this book explores correlations between the findings in developmental neurobiology and the behavioral effects of plasticity, stressing the relevance of findings in the biological sciences to the effects on cognitive functioning. Some tentative examples are provided, where data derived from developmental neurobiology may have relevance to the design of strategies for the remediation or enrichment of cognitive functions that make use of the plasticity of the immature brain. I hope that this will be of interest not only to students in the various disciplines involved, but also to educators and clinicians concerned with preschool and school-age educational programs, both for normal and for developmentally handicapped children, and to the many educated laypeople for whom knowledge concerning the development of the human brain and of its functions holds fascination.

# 1

## NEUROANATOMICAL SUBSTRATES: EARLY DEVELOPMENTAL EVENTS

> I noticed that every outgrowth, dendritic or axonic, in the course of formation, passes through a chaotic period, so to speak, a period of trials, during which there are sent out at random experimental conductors, most of which are destined to disappear.
>
> Santiago Ramon y Cajal, 1937

The fact that the environment of an organism affects its function has long been known. However, proof that even subtle changes in input from the outside world can affect the anatomy of the brain is relatively new. Environmental effects on brain anatomy are seen primarily in the infant and young child, but persist to some extent in the adult. Prior to birth, in the embryo and fetus, development of the cerebral cortex appears to be relatively fixed, with respect both to timing and to the nature of the developmental program. The effects of the external environment are as yet very limited. Factors in the internal environment may influence development of the cerebral cortex during the prenatal phases of brain development, but do so primarily in a negative fashion.

The early development of the cerebral cortex appears to be to a large extent genetically determined. It involves a complex series of sequential steps that are controlled by a large number of developmentally expressed genes (genes that are transiently activated to produce specific gene products during specific periods of development). A brief summary of these steps appears to be appropriate for this discussion of cortical plasticity, because it provides the background that leads up to the point at which environmental inputs are most likely to modify the developmental program. A more detailed description of the early steps of cortical development in the human may be found in the excellent atlas by O'Rahilly and Müller (1994) and in a review by Sidman and Rakic (1973).

### Neural Induction and Neural Tube Formation

The location of the cells that will make up the cerebral cortex is already defined in the late gastrula stage of the embryo, at about 18 postovulatory days (days since estimated last ovulation, POD) in the human, when the length of the embryo measures only about 1.5 millimeters. These cells are derived from the ectoderm (outer layer of cells) located near the midline, on the dorsal (back) surface of the embryo. The region that develops into the nervous system is at this stage represented by a midline groove (the neural groove), surrounded by a slight elevation, the neural plate (Figure 1.1A). The rostral (forward or anterior) portion of the neural plate develops into the cerebral cortex. At first, the ectodermal (outer or superficial) cell layer in this region is multipotential, able to develop into any part of the embryo, as has been shown by transplantation experiments in animals, where the dorsal ectoderm has been transplanted to other sites of the embryo. There follows an event referred to as neural induction, which specifies the cells for neural development. The process of neural induction has been studied extensively, but our understanding of it is still incomplete (Jacobson 1991, chap. 1). Cells that are located rostrally in the neural plate region become specified for cerebral cortical development; those located more caudally (toward the tail end) form the spinal cord. Signals received from cells underlying the neural ectoderm appear to be important for the specification that leads to development of specific portions of the neuraxis.

The next developmental step is a rounding up of the neural groove to form a tubular structure, the neural tube. Our interest is focused on the rostral portion of the neural tube, which develops into the cerebral cortex. Closure of the rostral end of the neural tube (the anterior neuropore) occurs at about POD 24 (Figure 1.1B). This is followed by division of the primitive forebrain (prosencephalon, Figure 1.1C) into two cerebral hemispheres and by subdivision of the central cavity of the rostral end of the neural tube into three compartments, the lateral ventricles and the midline third ventricle, at about POD 32, when the embryo measures about 7 millimeters (Figure 1.1D).

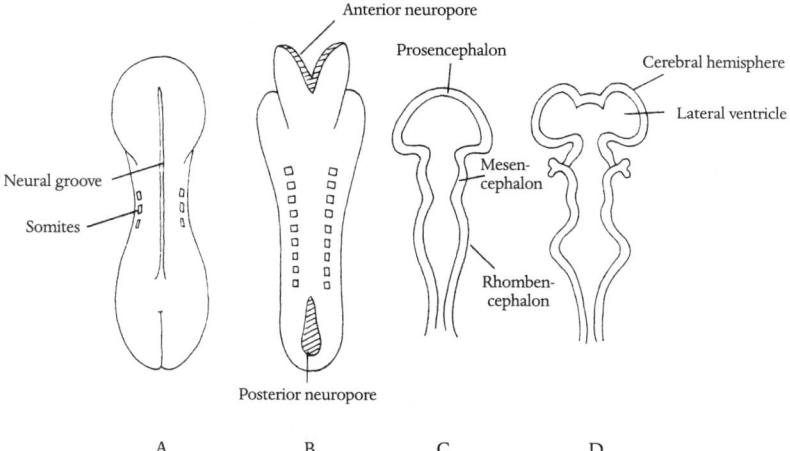

Figure 1.1. Early developmental stages of the human central nervous system. A: Dorsal view of the embryo at about the conceptual age of 20 days. The future nervous system is indicated by a midline depression, the neural groove. B: Dorsal view of the embryo at about the conceptual age of 23 days. The neural groove has closed dorsally to form a neural tube, except for openings at either end, the anterior and posterior neuropores. C: The cephalic portion of the embryo, conceptual age about 28 days. The future cerebral hemispheres are represented by a single midline expansion of the rostral end of the neural tube, the prosencephalon. D: The cephalic portion of the embryo, conceptual age about 36 days. Paired cerebral hemispheres and lateral ventricles are formed. (From Behrman and Vaughan, 1987, fig. 21-6, p. 1299; reprinted with the permission of the W. B. Saunders Company.)

## The Birth of Neurons

Following the development of the two lateral forebrain cavities, there is a remarkable proliferation of the neuroepithelial cells in this region. This process begins at about POD 52, still very early in embryonic development, when the length of the embryo measures about 24 millimeters. It occurs in a zone adjacent to the lateral ventricle known as the germinal matrix. This zone produces nearly all of the millions of neurons that make up the cerebral cortex. It also generates many of the supporting cells of the brain, the astroglia, and the cells necessary for the formation and maintenance of the myelin sheath, the oligodendroglia. The location of a dividing cell in the wall of the lateral ventricle determines its eventual position in the forebrain. Cells located ventrally form the basal ganglia; those located laterally and dorsally lead to formation of the cerebral cortex. As cell division progresses, some of the dividing cells become postmitotic (no longer able to undergo cell division), and mature into either glial (supporting) or neuronal cell types.

## Neuronal Migration

Some early glia grow processes that attach both to the ventricular lining and to the surface of the neural tube. These radial glia, discovered by Pasko Rakic (1971), form a scaffold along which postmitotic neurons migrate from the germinal matrix zone to the cortical plate (the region located under the surface of the neural tube). The great majority of migrating primitive neurons appear to move along the guides provided by the radial glia. There are, however, smaller populations of neurons that migrate together tangential to the radial glia fibers, and that may reach the cortical plate at a considerable distance from their point of origin (Walsh and Cepko, 1988).

Some of the earliest arriving cells set up transient cell populations below the cortical plate (see subplate neurons, Figure 1.2 below; also Kostovic and Rakic, 1990; Goodman and Shatz, 1993; Allendoerfer and Shatz, 1994). Others migrate to positions just above the cortical plate in the marginal layer (Cajal-Retzius cells, Figure 1.2). These two cell populations gradually disappear, but some Cajal-Retzius cells and subplate neurons persist until after birth. Cajal-Retzius cells have been reported to persist in human prefrontal cortex until early adult life (Mrzljak et al., 1990). Chun and Shatz (1989) have reported the persistence of some subplate neurons in adult feline subcortical white matter, as identified by early neuronal birth verified by 3H-thymidine labeling[1] and by characteristic immunohistochemical mark-

---

1. The tritiated thymidine (3H-thymidine) and bromodeoxyuridine (BdU) methods for dating of neuronal birth are mentioned repeatedly in this chapter. They are based on the fact that the genetic material (DNA) of a cell persists unchanged for the lifetime of the cell, unless further cell divisions occur. Since neurons are postmitotic cells, they do not undergo further cell divisions. A label introduced into the DNA at the time of the last cell division, which occurs in the embryo prior to neuronal migration, therefore persists for the lifetime of the cell. Neurons usually survive for the entire life span of the organism. A neuron radio-labeled by injection of the radioactive nucleotide 3H-thymidine or BdU in the embryo may therefore be identified in the adult animal as having been born (specifically having undergone mitosis and DNA synthesis) on about the day of the tracer injection.

ers.[2] Study of the subplate zone in human brains has shown subplate neurons to be maximally represented between the gestational ages (GA) of 22 and 34 weeks, with postnatal disappearance after the age of 6 months. Some of these neurons may survive in adult human cerebral cortex (Kostovic, Lukinovic, Judas, et al., 1989).

One of the functions of these two early neuronal populations appears to be that of providing signals or guideposts for migrating neurons and/or growing axons (Clark et al., 1997). Cajal-Retzius cells may be important for determining the point at which migrating neurons detach from radial glia and assume their position in the cortical plate, which is initially just below the Cajal-Retzius cells of the marginal layer (later termed molecular layer or layer 1 of the cerebral cortex). Neurons never migrate past the Cajal-Retzius cells into the marginal (molecular) layer. Subplate neurons may be important for the establishment of the correct afferent pathways to the cortex, for example for the guidance of axons that arise from the lateral geniculate nucleus and that terminate in layer 4 of the visual cortex (Ghosh, Antonini, McConnell, and Shatz, 1990). The disappearance of these transient neuronal populations is an example of what has been referred to as programmed cell death, or apoptosis, a phenomenon that is widespread in the development of the nervous system, and that is important for the elimination of redundant cell populations (Cowan et al., 1984).

The migration of immature neurons has been studied in animal models using radio-labeled (tritiated or 3H-) thymidine as a marker (Sidman and Rakic, 1973). This radioactive compound permanently labels the DNA in the nuclei of neurons that are born (become postmitotic) shortly after injection of the tracer. It can therefore be used to determine the eventual location in the cerebral cortex of cell populations that are born at a given time. This method has shown an inside-out pattern of cortical development: Neurons in the lower layers of the cerebral cortex arrive first. Neurons born later migrate past the early arrivals and form the upper layers of the cerebral cortex. In that way, newly arriving neurons always assume a position just below the molecular layer, consistent with a signaling effect by the

---

2. Different cell types have specific cell surface proteins, which can be demonstrated by a label, such as a dye, attached to an antibody to the cell surface protein. The labeled cells can then be visualized in histologic sections.

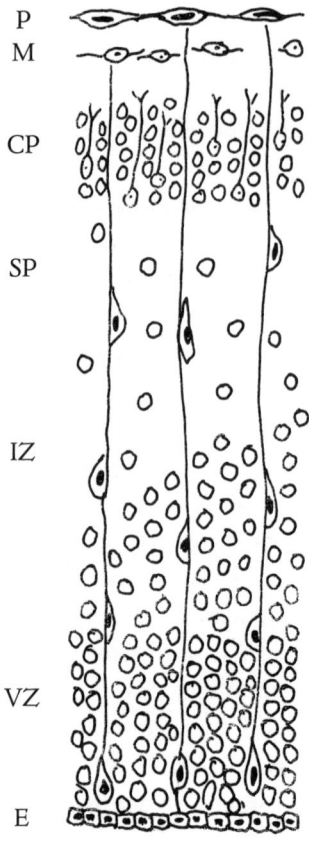

P
M
CP
SP
IZ
VZ
E

Figure 1.2. The cerebral wall of the human embryo at about the conceptual age of 110–120 days. P = the pia, or outer limiting membrane of the cortical plate. M = the molecular layer, in which Cajal-Retzius cells are located; these are transiently present neurons that may be important for signaling the end of neuronal migration. CP = the cortical plate, where neuronal migration ends and cerebral cortex with distinct layers is formed in an inside-out fashion (neurons forming the lower cortical layers arrive first, and later-arriving neurons migrate past them to form the upper cortical layers). SP = the subplate zone, a relatively acellular area that contains the subplate neurons which act as guideposts for axons of thalamic neurons growing into the cortical plate. IZ = the intermediate zone; vertical migration of neurons along glial fibers and tangential migration independent of glial guides occur in this region. VZ = the ventricular zone (the germinal epithelium), the region of proliferation of neuronal and glial precursor cells. E = the ependyma (lining of the lateral ventricle). (Drawing by Arun S. Dabholkar.)

overlying Cajal-Retzius cells of layer 1. Recently, platelet activating factor (PAF), which is present in embryonic Cajal-Retzius cells, has been identified as a possible signaling molecule for detachment of migrating neurons from the radial glia (Clark et al., 1997). The time course of the development of cortical neurons, including the growth of the cortical dendrites, recapitulates that of neuronal migration. The neurons in the lower cortical layers, which arrive first at the cortical plate, develop earlier than the neurons in the upper layers (Marin-Padilla, 1970).

At the age when the human fetus becomes viable, that is to say at about postovulatory(PO) week 25, most neurons have formed and neuronal migration in the cerebral cortex is nearly complete. There is, however, some persistence of neurons in the subcortical white matter, probably late migrating stragglers, as well as survivors of the primitive embryonic subplate neurons (Figure 1.2).

The major events related to prenatal neurogenesis of the cerebral cortex are under genetic control. Genes necessary for normal development of the cerebral cortex have recently been discovered. Mutations in these developmentally important genes lead to developmental malformations, which are major causes of mental retardation, cerebral palsy, and epilepsy. For example, the division of the primitive forebrain (the prosencephalon) into two cerebral hemispheres (Figure 1.1C) depends on the action of several genes. Defects in these genes may lead to a brain whose development is arrested at the unpaired forebrain stage (holoprosencephaly)(Ming and Muenke, 1998). Other midline defects, such as the presence of a single midline eye (cyclopia), are also seen. Several genes are known to be necessary for normal formation of the paired cerebral hemispheres. These genes are partly conserved in all species from the fruit fly *(Drosophila)* to the human. They are referred to as homeobox genes. Some have been given fanciful names such as "sonic hedgehog." Multiple abnormalities in the sonic hedgehog gene on chromosome 7q have been associated with holoprosencephaly in humans (Roessler and Muenke, 1998). Recently, another homeobox gene, on chromosome 2p, has been found to also be important for normal formation of paired forebrain structures (Wallis et al., 1999).

A number of genes are critical for normal migration of neurons. Abnormalities in these genes are associated with migratory defects in which large masses of neurons, normally destined to form the cerebral cortex, become arrested, either in the subependymal region near the germinal matrix or in the subcortical white matter. In the most severe forms of migration defect that are compatible with survival of the neonate, migration of the majority of neurons becomes arrested in the region of the subcortical white matter. Not enough neurons reach the cortical plate to form the cortical convolutions. The surface of the brain is therefore smooth (a condition called lissencephaly). One form of lissencephaly is due to a mutation in a gene on chromosome 17 (the LIS-1 gene), which codes for a subunit of the platelet activating factor acetyl hydrolase. The gene products of LIS-1 have been found to be highly concentrated in Cajal-Retzius cells during the period of neuronal migration (Clark et al., 1997). Mice with mutations in the LIS-1 gene have had migratory defects, with a severity approximately proportional to the degree of deficiency of the LIS-1 gene product (Hirotsune et al., 1998). These studies suggest that the gene product of the LIS-1 gene may be a signaling factor for neuronal migration.

A human migratory defect in which populations of neurons fail to migrate altogether, and form nodular masses in the periventricular zone, has recently been found to be due to point mutations in the filamin 1 gene (Fox et al., 1998). Filamin 1 is an actin-binding protein that mediates changes in cell shape that are important for the migration of many cell types. In the central nervous system, it is developmentally expressed in neurons during the period of migration into the cerebral cortex. As soon as they reach their adult position in the cerebral cortex, neurons lose this protein and at the same time lose the ability to move.

### Environmental Effects on Early Developmental Events

Changes in the internal environment, as well as genetic defects, may influence the early developmental steps. Environmental effects in the embryo and fetus have mainly negative consequences. A partial list of such pregnancy risk factors is provided in Table 1.1. The list is so long that it appears miraculous that most children turn out to be normal. There is little evidence that anything women do other than practicing good nutrition and avoiding the risk factors listed in Table 1.1 will substantially improve pregnancy outcomes as far as the structure or the function of the cerebral cortex is concerned. Recently, attempts have been made to provide "environmental stimulation" to the fetus in utero, in the hope that this will improve the child's ability to learn later on. While playing music to the fetus in utero appears to decrease fetal movements, there is to date no convincing evidence that such stimulation will enhance cognitive functions later in life. A study by Lafuente and colleagues (1997) reports enhanced psychomotor development at the age of 6 months in infants exposed to music by Mozart during their fetal life. However, this study suffers from failure to consider a possible reporting bias on the part of the mothers in the experimental group. Some data suggest that excessive sound and light stimulation of the prematurely born infant may have negative effects and should be avoided. The present practice in premature infant nurseries is to keep the rooms dimly lit and as quiet as possible.

Negative effects of the environment on the development of the cerebral cortex have been investigated extensively. These effects range from com-

*Table 1.1.* Negative effects of environmental factors on brain development.

| Risk factor | References |
| --- | --- |
| Maternal drug and other toxic exposure | |
| Cocaine | Gressens, Kosofsky, and Evrard, 1992 |
| Alcohol | Gressens et al., 1992; Miller, 1986; Miller and Potempa, 1990; Miller and Robertson, 1993 |
| Cigarette smoking | Fedrick and Anderson, 1976 |
| Toluene inhalation | Arnold et al., 1994; Gospe and Zhou, 1998; Pearson et al., 1994 |
| Valproic acid, phenytoin, folic acid antagonists | Friedman and Polifka, 1994; Koch et al., 1992 |
| Fetal and neonatal infection | Greenfield, 1958; Crome and Stern, 1972 |
| Cytomegalic inclusion disease (CMV) | |
| Herpes simplex virus | |
| Toxoplasmosis | |
| Acquired immunodeficiency disease | |
| Nutritional deficiency | |
| Vitamin and mineral deficiencies | Cravioto and Delicardie, 1970; Winick, 1970 |
| Folic acid | MRC Vitamin Study Research Group, 1991 |
| Pyridoxine (vitamin B6) | Czeizel and Dudas, 1992 |
| Iron deficiency. | Oski, 1993 |
| Hormonal abnormalities | Eayrs, 1971; Nicholson and Altman, 1972 |
| Deficiency of thyroid hormone | Rami et al., 1986a,b |
| Corticosteroid hormone excess. | Gumbinas, Oda, and Huttenlocher, 1973; Lampe, Touch, and Spitzer, 1999 |
| Ionizing radiation, including exposure to x-rays | Schull, 1999; Schull and Otake, 1998 |
| Inborn metabolic errors | |
| Phenylketonuria (PKU) | Bauman and Kemper, 1982 |

plete failure of the forebrain to develop to subtle effects on the fine structure of the brain and on cognitive functions. For example, folic acid deficiency in the first few weeks of pregnancy leads to an increased incidence of failure of the neural tube to close. This results in spina bifida if the caudal end of the tube fails to close, in anencephaly (absence of forebrain structures, the most severe of cortical malformations) if the closure defect involves the rostral end of the tube, or in rachischisis, a closure defect of the entire neurotube that includes anencephaly plus spina bifida. In humans, folic acid supplementation, started prior to conception, has markedly decreased the incidence of these severe developmental defects (Czeizel and Dudas, 1992; Berry et al., 1999).

Exposure to ionizing radiation at remarkably low levels results in a small brain (microcephaly) due to a decrease in the number of neurons that are formed, if it occurs during the period of neuronal proliferation, between about POD 52 and PO week 20. Extensive data concerning the susceptibility of the human fetus to ionizing radiation derive from the study of the offspring of pregnant women exposed to radiation from the atomic bomb explosions in Hiroshima and Nagasaki in 1945 (see below).

Several other risk factors, listed in Table 1.1, may also lead to microcephaly. This malformation is, in turn, a major cause of mental retardation. The developing brain is, however, relatively well protected from damage by many environmental factors. For example, maternal-fetal malnutrition will lead to a smaller body size and weight of the infant, and will increase the incidence of infant mortality. The development of the brain, however, tends to be spared both in animals and in malnourished human populations. A well-known study involved the follow-up of infants born during the Dutch "hunger winter" of 1944–45, when the German occupation forces starved the population in a large part of the Netherlands as punishment for collaborating with the Allies. Infants born to starved mothers were compared to a cohort without a history of severe nutritional deprivation. Both groups were studied at the time of military induction, specifically at the age of 19 years. No effects of fetal malnutrition on brain size or function were found in this study, although the infants born during the starvation period had significantly decreased body weights at birth, as well as increased infant mortality (Stein et al., 1972, 1975).

In contrast, studies on the effects of postnatal malnutrition on brain development have reported significant effects of severe malnutrition in the first postnatal year, especially when protein-calorie deprivation occurred over a period of more than 4 months. The observed effects included microcephaly, cognitive impairments, and abnormal behavior (Winick, 1970; Cravioto, 1970). These studies have often been carried out in rural areas of underdeveloped countries. In such settings, it has been difficult to separate the effects of protein-calorie malnutrition on brain development from those on muscular development, since infant tests of cognitive ability often require motor responses that may be delayed in infants with underdevelopment of muscles due to malnutrition. The cognitive functions of malnourished children may also be negatively affected by specific nutritional deficiencies, such as iron deficiency (Oski, 1993), by emotional deprivation, neglect, and disease, especially viral infections.

Drugs ingested by the mother may affect fetal brain development, but the effects in humans may be difficult to separate from other factors related to drug use during pregnancy, including the poor health of the mother, especially when there is maternal drug addiction. Infants exposed to maternal cocaine use during pregnancy may have pre- or perinatal stroke due to a vasoconstrictive effect of cocaine, which may lead to vascular occlusion. Animal studies have in addition shown diffuse effects of cocaine on the development of the cerebral cortex, with defects in the formation of cortical layers (Gressens, Kosofsky, and Evrard, 1992).

Exposure of the fetus to high alcohol levels inhibits brain growth (Miller, 1986; Miller and Potempa, 1990; Miller and Robertson, 1993), resulting in small adult brain size and cognitive deficits. In the human, there are in addition specific facial features, nail changes, and cardiac defects, making it possible to recognize a "fetal alcohol syndrome" (Jones and Smith, 1973). Abuse of inhalants from commercially available spray cans during early pregnancy has led to abnormal brain development, including microcephaly and mental retardation (Arnold et al., 1994). The toxic component appears to be toluene. When administered to pregnant rats, this organic solvent has also been shown to lead to defective genesis of neurons and abnormal neuronal migration in the offspring (Gospe and Zhou, 1998).

Recently, there has been concern over the possible negative effects of maternal corticosteroid hormone use on the neurodevelopment of the offspring (Lampe et al., 1999). Prenatal exposure to pharmacologic doses of corticosteroids, administered to promote fetal lung maturation, has been reported to be associated with small head size and body size (French et al., 1999) and with specific cognitive defects, including a decrease in visual memory (McArthur et al., 1982). The last is especially interesting, since it suggests possible impairment of hippocampal functions. Impaired development of the hippocampus, including dentate gyrus, has been found in rhesus monkeys after maternal corticosteroid hormone administration (Uno et al., 1990). Hippocampal neurons may be especially vulnerable to injury by corticosteroid hormones because they have a high level of cortisol receptors.

When environmental effects on fetal brain development have been noted, they have often consisted of a reduction in brain size, suggestive of a reduction in the number of neurons, diminished dendritic growth, or both. The effect on the developing brain depends to a large extent on the timing of exposure to the toxic agent or teratogen. For example, folic acid deficiency leads to anencephaly only if it exists prior to and up to the time of normal closure of the neural tube, that is, about POD 24. This brief time period represents the "critical period" for neural tube closure defects. Critical periods occur throughout the development of the central nervous system, and they differ greatly for different developmental events. They are time windows during which a specific developmental process is particularly sensitive to perturbation by environmental factors, usually at the time of rapid growth. Such perturbations often have negative effects, such as in the example of folic acid deficiency. This effect has given rise to the concept of "vulnerable periods." However, as will be pointed out later, some critical periods may also be exploited for positive effects on brain structure and function. This type of critical period effect has been referred to as a "window of opportunity."

The concept of critical periods in development was first articulated by John Beard in 1896. In a Harvey Lecture, L. B. Flexner (1951–52) defined a critical period in the development of the brain of the guinea pig fetus, during which there is sudden anatomic, biochemical, and functional maturation of the brain. Dobbing (1981) stressed the sensitivity of rapidly developing neural structures to damage by negative environmental factors, including metabolic abnormalities and cerebral hypoxia ("vulnerable periods"). The third trimester of pregnancy and early infancy appears to be such a vulnerable period for the human, probably because of the rapid growth of the neuropil, including axons, dendrites, and synapses, that occurs during this period. Dobbing refers to it as the "brain growth spurt" (Davison and Dobbing, 1968; Dobbing and Sands, 1979). At the same time, it has become clear that it is not possible to define a single critical period for human brain development. Critical periods are more clearly definable in the development of simple brains, such as that of the guinea pig. In human brain development there are multiple critical or vulnerable periods, depending on the developmental event. Thus there are critical periods for neuron formation, for neuronal migration, and for dendritic and synaptic development, and they differ markedly, one from the other.

Dobbing and Sands (1970, 1973) defined the period during which the birth of most of the neurons occurs in the human forebrain. The method used was determination of the total DNA in the forebrain at various conceptual ages. Postmitotic cells maintain a constant amount of DNA per cell. The value for total DNA therefore is a measure of cell number. Using this method, researchers have found that the birth of cortical neurons, while it has its onset at about POD 52, is greatest between about POD 70 and 140. Proliferation of cells continues in the forebrain at a slower rate after POD 140, but most of these late-forming cells are probably glial rather than neuronal. The occurrence of neurogenesis between about POD 52 and 140 agrees well with the period of increased sensitivity of the fetus to ionizing radiation, which is known to damage rapidly dividing cells. Information concerning the time window for maximum radiation damage to the developing brain comes largely from the study of pregnancy outcomes in women exposed to gamma radiation from the atomic bomb explosions in Hiroshima and Nagasaki in 1945. These data indicate that the period from POD 56 to 105 is one of special sensitivity to radiation-related brain damage, almost exactly the period of neuronal proliferation in the forebrain found by Dobbing and Sands. The effects seen in the radiation-exposed infants included microcephaly and mental retardation, which are the expected outcomes of reduced neuronal number in the forebrain (Schull, 1998; Schull and Otake, 1999).

Neuronal migration occurs immediately after the birth of neurons, especially during POD 110–140. This period might be expected to be one of sensitivity to migratory defects in the forebrain. Indeed, defects in the formation of the cerebral cortex, especially the formation of a cortex that contains numerous small gyri (polymicrogyria), have been reported as a result of maternal carbon monoxide poisoning and of intrauterine infections during this time period (Greenfield 1958, p. 335; Crome and Stern, 1972, pp. 39, 49).

In the human, there is an important critical or vulnerable period that extends from the newborn period up to at least the age of 2 years, and that coincides with the formation of dendritic and axonal branches and synaptic contacts. Myelination of the subcortical white matter also is at its maximum rate during this time. Hormonal disorders, especially thyroid hor-

*Table 1.2.* Critical (vulnerable) periods in the development of human cerebral cortex.

| Malformation | Critical period |
| --- | --- |
| Anencephaly | POD 7–25 days |
| Microcephaly (decreased neuronal number) | POD 52 to end of second trimester |
| Migratory defects | Second trimester |
| Defective formation of dendrites and synapses | Third trimester and first postnatal year |

mone deficiency, and a number of inborn metabolic errors, most notably phenylketonuria (PKU), lead to permanent stunting of axonal and dendritic growth, to abnormal synaptic development, and to mental retardation unless the defect is corrected prior to the critical period. This fact has led to neonatal screening programs that diagnose the deficit at birth, making it possible to prevent mental retardation in a significant number of children. A summary of human critical periods in brain development is provided in Table 1.2.

### The Formation and Migration of Cortical Neurons in the Adult Brain

There are several areas in the primate brain where formation of neurons continues after birth. These include the dentate gyrus in the hippocampus and the olfactory neurons (Rousselot et al., 1995; Doetsch et al., 1997; Gould and Tanapat, 1997; Eriksson et al., 1998; Gould, Beylin, et al., 1999; Gould, Reeves, Fallah, et al., 1999). The study by Eriksson et al. (1998), carried out in postmortem human brains, yielded results similar to those found in studies of subhuman primates. Recently, evidence has been presented that formation of new neurons also occurs in adult *neocortex*. Neuronal birth in the subependymal zone, migration through the white matter, and incorporation into the neocortex have been described in primates during adult life. Adult formation of new neurons has been reported in three cortical association areas, prefrontal, inferior temporal, and parietal cortex (Gould, Reeves, et al., 1999), but not in primary sensory (calcarine) cortex. Evidence derives from bromodeoxyuridine (BdU) labeling of cells that have undergone recent mitosis. Such cells are seen in the subventricular zone of adult monkeys within hours of BdU injection, in the subcortical white matter 1 week after the injection, and in cerebral cortex at 2 weeks. Immunohistochemical markers have been used to identify these cells as neurons. The total number of cells so identified has been small. The method requires strict specificity of staining by neuronal and glial markers, to make certain that the cells are not glia.

Prior studies on neurogenesis, using 3H-thymidine autoradiography, had failed to show adult formation of new neurons in the neocortex of primates (Rakic 1985, 1998). In these studies, the time of survival after injection of 3H-thymidine was much longer than that in the study by Gould, Reeves, et al. (1999). The difference in results therefore may be related to the short survival time of late-forming neurons. Contrary to neurons that form during fetal life, hippocampal neurons that form in the adult brain tend to disappear after some weeks (Kempermann et al., 1997; Gould, Beylin, et al., 1999). The survival time of the neocortical neurons that form in adults is as yet unknown. Differences in results may also have been related to the region of neocortex that was studied. No evidence of adult neurogenesis was found by Gould and colleagues in the visual cortex, the area studied earlier by Rakic. At any rate, the formation of new neurons in the adult must be approximately balanced by the loss of neurons, since there is no evidence for progressive increase in the size of the cerebral cortex during the adult years.

An interesting question is whether subjecting animals or humans to learning and memory tasks can prolong the survival of late-forming cells, that is, whether these cells are available for incorporation into functioning systems when there is increased demand for information processing in a cortical system. Recent evidence suggests that prolonged survival of late-forming hippocampal neurons may occur under certain conditions (Kempermann et al., 1997; Gould, Beylin, et al., 1999). These findings raise interesting questions. Are late-forming neurons important for adult neural plasticity? Is formation of new neurons a mechanism for recovery after focal brain damage, both in the adult and in the child? Can the rate of formation of these cells be increased by manipulation of the regulatory system that normally results in the turning off of developmental events in the cerebral cortex? If so, the induction of neuronal proliferation might become an approach to therapy for a large number of neurodevelopmental disorders, including microcephaly.

The occurrence of adult neuron formation in cortical association areas, and not in the primary sensory cortex, indicates that care needs to be taken not to generalize findings in one cortical region to the cortex as a whole.

Differences appear to exist in developmental patterns. The formation of neurons in the association cortex in the adult, and the absence of these in the visual cortex, may be one of the anatomical bases of greater neural plasticity in the association cortex.

### Dendritic Development

After arrival at their permanent location in the cortical plate, neurons begin elaboration of their dendritic trees and outgrowth of their axons. This almost immediately leads to synapse formation, an event that will be discussed in more detail, since it is a developmental step which—especially in its later phases—is sensitive to environmental changes, both internal, such as those produced by focal brain lesions, and external, those related to changes in sensory input.

Mature cortical neurons have extremely complex shapes, including multiple dendritic branches arising from the cell body that in turn have secondary branchings, and an axon that emerges from the cell body, and then divides into multiple branches. Some axons may be remarkably long, up to 1 meter in length in the case of the giant Betz cells in the motor cortex. The dendritic and axonal branches are essential components of the signaling systems that underlie cortical functions. The dendrites collect information from other cells in the form of local changes (depolarization or hyperpolarization) of the cell membrane. These are mediated by synapses between axon terminals and specialized postsynaptic membranes on the dendrites. When local membrane depolarizations reach a critical threshold level, they lead to activation of the entire cell, and to a wave of depolarization of the axon, which transmits information to other (target) cells.

The dendritic branches are especially well shown by the Golgi method. This amazing histologic method, discovered serendipitously by Emilio Golgi (1903), stains neurons and their processes completely, but does so in only about 1 in 50 neurons, apparently on a random basis. This makes it possible to visualize the dendritic tree of individual neurons and to study the three-dimensional structure of neurons in thick sections (usually 100 or more microns). The shape of neurons in experimental animals can also be studied by injection of single cells in vivo with supravital dyes via a micropipette inserted into the neuronal cell body. The dye is then trans-

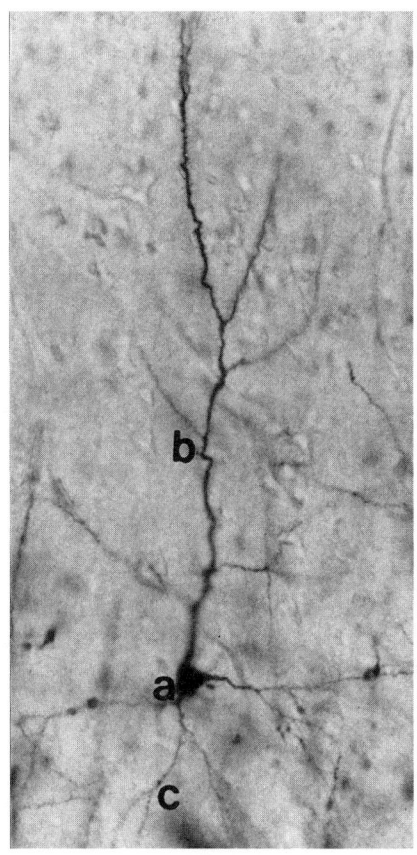

Figure 1.3. A typical pyramidal neuron in human neocortex, shown by the Golgi-Cox method: a = cell body; b = apical dendrite; c = basal dendrite.

ported by the living cell into all dendritic branches. This recent method has to a large extent validated the results obtained with the Golgi method.

The most common and characteristic type of neuron in mammalian cerebral cortex is the pyramidal cell (Figure 1.3). It has a pyramidal-shaped cell body, a large vertically directed process that arises from the cell body and extends toward the surface of the cerebral cortex (the main apical dendrite), and usually a single axon that arises from the base of the cell body, and then divides into multiple branches. The cell body also gives rise to an average of about six basal dendritic branches, each of which has secondary and tertiary branchings. The main apical dendrite also has branches, which in turn form secondary and tertiary branchings. A large pyramidal cell in human neocortex may have lateral dendritic branches that extend for a distance of 1 or more millimeters. Axonal branches may be much longer. In the case of the large pyramidal cells of the motor cortex (Betz cells) they extend all the way to the spinal cord. Other cell shapes also occur: stellate (multipolar) cells, which lack a main apical dendrite and which extend dendritic branches equally in all directions, and bipolar cells, which have two main dendrites, one pointing upward to the cortical surface, the other pointing down toward the white matter. The latter are most common in the hippocampal gyrus.

A universal feature of neurons stained by the Golgi method is the presence of many small excrescences (outgrowths) on the dendrites, the dendritic spines. These are regions in which dendrites make synaptic contact with axonal branches. Dendritic spines therefore are markers by which the presence of synapses can be assessed at the light microscopic level. Study of dendritic spines during development has shown marked reshaping of the connections. For example, dendritic spines are common on the cell body and the initial dendritic segment of developing neurons, but disappear from these locations during maturation (Larramendi, 1969). In addition, the density of spines on the apical dendrite has been reported to decrease during development. Such a decrease in spine density has been found in human visual cortex during the first postnatal year (Michel and Garey, 1984). Decrease in spine density on dendrites may be due to loss of synapses ("synaptic pruning"). It may also be accounted for by dendritic growth: elongation of a dendrite without change in the total synapse number will lead to a decrease in the density of synapses on the dendrite. Becker and colleagues (1984) have published a quantitative Golgi study of developing human visual cortex. It shows that dendritic growth indeed occurs during the early postnatal period, at the time when dendritic spine density decreases. It is likely that both processes—the pruning of synapses and decreased spine density related to elongation of the dendrites—occur during postnatal development of the cerebral cortex. Evidence of synaptic pruning derives from electron microscopic studies, which will be discussed in Chapter 2. The fact that synapses are much more densely packed on the dendrites of cerebral cortical neurons of infants and young children than on those of adults may account for the increased excitability of the cerebral cortex in the young child. An example is the tendency for fever to lead to convulsions ("febrile seizures"), a condition seen between ages 3 months and 5 years. It

*Table 1.3.* Mean dendritic length of cortical pyramidal neurons as measured in Golgi-stained sections at various ages. MFG = middle frontal gyrus; Calc = calcarine cortex.

| Area | Birth | 6 months | 24 months | 7 years | Adult |
|---|---|---|---|---|---|
| MFG, layer 3[a] | 203 (3)[c] | 2,369 (35) | 3,259 (48) | — | 6,836 (100) |
| Calc, layer 3[b] | 950 (33) | 1,150 (40) | 2,900 (100) | 2,400 (83)[d] | — |
| MFG, layer 5[a] | 858 (11) | 4,246 (56) | 4,669 (62) | — | 7,558 (100) |
| Calc, layer 5[b] | 1,250 (42) | 2,100 (70) | 2,800 (93) | 3,000 (100)[d] | — |

a. Data for the MFG are from Schade and van Groenigen, 1961.

b. Data for the calcarine cortex (Calc) are from Becker et al., 1984, fig. 3A-3D. Approximate values for apical dendrites (Becker et al., 1984, fig. 3A, 3B) have been added to values for basal dendrites (Becker et al., 1984, fig. 3C, 3D) for estimates of total dendritic length.

c. Dendritic length values are in micrometers. Numbers in parentheses indicate the percentage of dendritic length at the maximum.

d. Age 7 years was the oldest age examined, and adult data were not available in the study of Becker et al., 1984.

also is likely to be the main cause of the increased metabolic activity of the immature cerebral cortex (see Chapter 3).

At present, the data on dendritic development of the cerebral cortex in humans are very limited. These data are derived from the study of autopsy tissue by the Golgi method. Conel (1939–1963) studied the development of multiple cortical areas from PO 40 weeks to maturity. His published drawings of representative sections stained by the Golgi method show regional differences at PO 40 weeks and at the postnatal age of 1 month. For example, dendritic development appears less advanced in the prefrontal cortex than in the superior temporal gyrus near the primary auditory cortex at these two ages. At later ages, the complexity of dendritic branching becomes too great for judgments to be made on the basis of inspection of the published photographs. Unfortunately, Conel did not use any methods for quantitation of dendritic branching and length. Quantitative data are available for the middle frontal gyrus in the prefrontal cortex (Schade and van Groenigen, 1961; Schade, Van Backer, and Colon, 1964; Mrzljak et al., 1990; Koenderink and Uylings, 1996; Uylings, 2000) and for the visual cortex (area 17) (Becker et al., 1984). The data in the Schade and van Groenigen and in the Becker et al. studies are presented in such a way as to allow for some direct comparisons (Table 1.3). Differences are seen in the time course of dendritic growth in these two regions, suggesting that dendritic growth is heterochronous (occurs at different times) in different cortical regions. At birth, the mean dendritic length of the cortical pyramidal neurons in layer 3 of the middle frontal gyrus is only about 3 percent of the adult length. In layer 5 neurons, the mean dendritic length is 11 percent of the adult length. In contrast, layer 3 pyramidal neurons in the visual cortex already have 33 percent of the maximum dendritic length; layer 5 neurons have 42 percent. By the age of 24 months, the mean total dendritic length of layer 3 pyramidal neurons in the prefrontal cortex is only about half of the maximum, while in the visual cortex the maximum appears to have been reached by that age.

Study of aging human prefrontal cortex has shown that dendritic elongation occurs in this cortical region even in old age, perhaps as compensation for neuronal loss (Buell and Coleman, 1981).

The admittedly limited data on dendritic length show one other important difference between calcarine cortex and middle frontal gyrus (MFG): pyramidal cells in the prefrontal cortex are much larger than those in the primary visual cortex. The maximum observed dendritic length in MFG pyramidal cells is more than twice that in calcarine cortex. The smaller cell size accounts for the denser packing of neurons that is found in the visual cortex. It probably also relates to the relatively brief developmental period of the visual cortex. In other words, more simple neural systems have smaller neurons that take less time to develop. A unique feature of human cerebral cortex is the large size of cortical pyramidal neurons. This feature may be one of the factors that account for the slow development of the human brain, compared with that of all other species, including other primates. In general, the size of the cortical pyramidal neurons is greatest in the frontal cortex, and increases as one moves up the phylogenetic scale (Ramon y Cajal, 1960).

### The Outgrowth of Axons

The presence of growing axons, with specialized leading edges known as growth cones, is a prominent feature of human neocortex in the neonatal period. The outgrowth of axons is accompanied by considerable remodeling, as was noted by the neuroanatomist Ramon y Cajal early in the twentieth century (Ramon y Cajal, 1937). The quote at the beginning of this chapter sums up his findings. Whole systems of axonal connection may be lost. These include the callosal connections of the visual cortex in the cat (Innocenti, 1981, 1985; Innocenti, Fiore, and Caminiti, 1977) and fetal cortico-spinal axons arising from neurons in the occipital cortex in rats (Stanfield and O'Leary, 1985b). In other systems one sees developmental reduction of axons rather than complete loss. The developing optic nerves represent a group of axons in which there is significant developmental reduction (Rakic and Riley, 1983), as does the corpus callosum of the rhesus monkey (LaMantia and Rakic, 1990). The loss of axons during development is an example of a general principle that governs the late developmental events in the cerebral cortex, specifically those occurring peri- or postnatally. Late-developing structures, including dendritic trees (Greenough and Chang, 1988), axons, and synapses tend to go through a period of exuberant growth followed by elimination or trimming away of the excess. This will be discussed in more detail in Chapter 2.

The plasticity of axonal growth during prenatal and early postnatal development has been demonstrated especially well in the reaction of the developing brain to focal damage. For example, damage to the immature motor cortex in one hemisphere leads to the growth of axons from the opposite normal motor cortex to the ipsilateral spinal cord, which has lost its cortical innervation. Regrowth of pyramidal tract axons occurs after sectioning of the pyramidal tract in neonatal rodents (Kalil and Reh, 1979, 1982). No such compensation occurs in the adult (see Chapter 4). It has recently been reported that axonal growth in the adult is inhibited by myelin-associated axonal growth inhibitors (Z'Graggen et al., 1998.) These normally act to prevent the formation of excessive axonal growth, but they also prevent the regrowth of axons after they have been severed. In this respect, axons in the central nervous system differ markedly from those in peripheral nerves: peripheral axons have a considerable capacity for growth after transsection, including correct target finding, even in adults. The outgrowth of central nervous system axons after focal brain injury in adult rats can be induced by injection of antibodies to the myelin-associated axonal growth inhibitors (Z'Graggen et al., 1998). This discovery has wide-ranging implications regarding critical periods for neural plasticity. It suggests that these periods could potentially be experimentally manipulated. It may therefore become possible to regenerate damaged axons in the adult central nervous system. This work points the way for a possible approach to therapy for focal brain and spinal cord lesions in the adult as well as in the child.

Neurons are arranged in the cerebral cortex in layers that differ in the predominance of different cell types (Figure 1.4). Layer 1 (the molecular layer) has few neurons in the adult, but contains transient neurons, the Cajal-Retzius cells mentioned above, during development. Layer 2 has predominantly small pyramidal neurons. Layer 3 contains many large pyramidal cells. Layer 4 has a predominance of small multipolar cells. Layer 5 is another layer of large pyramidal cells, which include the giant Betz cells of the motor cortex. Layer 6 is pleomorphic, gradually merging into the subcortical white matter (Figure 1.4).

The different cortical layers subserve different functions: layer 4 receives afferent input from sense organs via the thalamus and the thalamocortical afferent fibers. For example, the axons of neurons located in the lateral geniculate nucleus of the thalamus project to layer 4 of the calcarine cortex. Layers 2 and 3 are concerned with processing the information received by layer 4, or received via connections from other cortical areas. They in turn have efferent connections to other cortical areas and to layer 5. Layer 5 contains important efferent systems of the cerebral cortex. It sends axons to other cortical or subcortical regions, sometimes over remarkably long distances.

While there are differences in the size of the layers in different cortical regions, an overriding impression is that of basic uniformity of the structure of the neocortex, as pointed out by Rockel and colleagues (1980). The cerebral cortex has a vertical, columnar organization, as was reported by Lorente de No (1938). These columns, first discovered as structural units, are now known to represent functional modules as well (see the discussion in Chapter 4 of ocular dominance columns and the columnar organization of the somatosensory cortex). They have been demonstrated in all mammalian species, where the basic structure of the columns is remarkably similar from mammals with simple cortical architecture, such as rodents, to man. The number of neurons per column is the same across species, although the volume of each column is much larger in humans than in rodents. Larger cell bodies and an increase in axons, dendrites, and synapses (neuropil) take up the increased space in humans.

Cortical regions with vastly different functions have the same basic structure, consisting of endlessly repeating vertically oriented groups of neurons

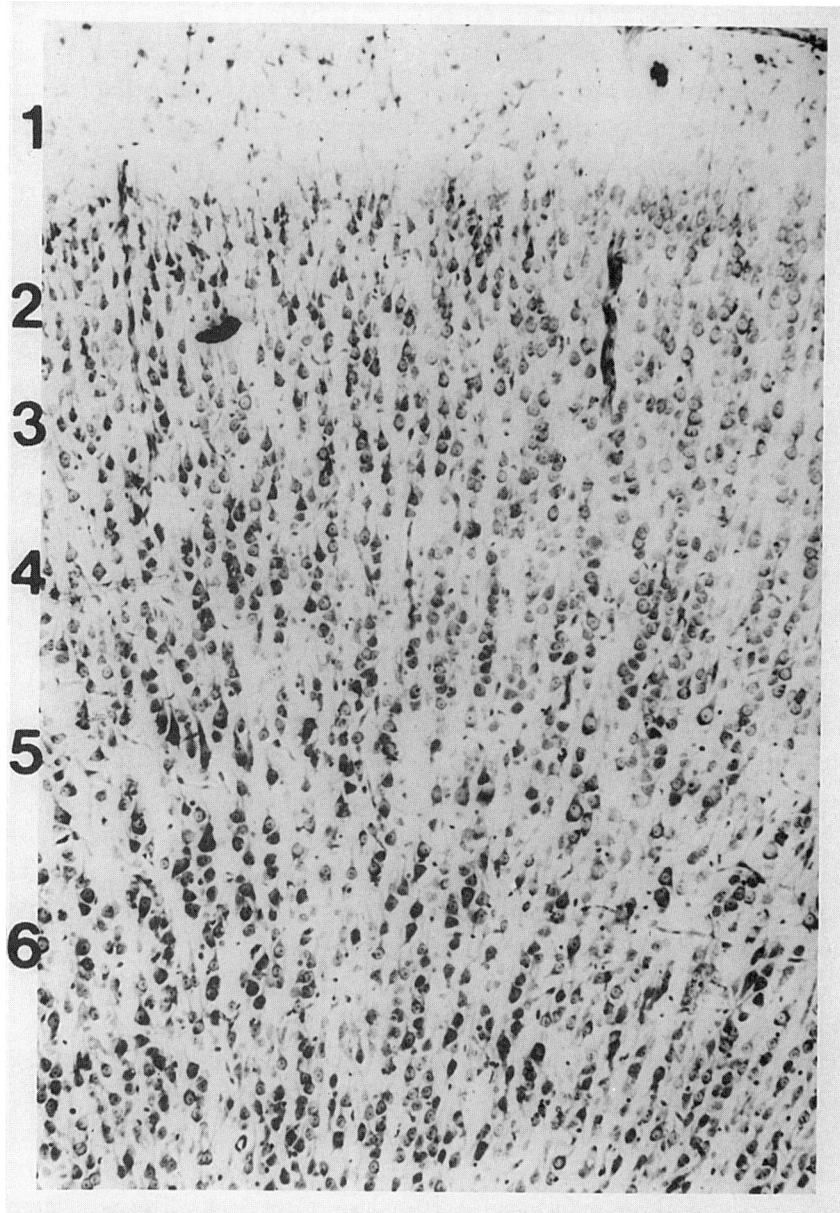

Figure 1.4. Typical six-layered neocortex. The layers are indicated by the numbers 1–6. This is a section of immature cortex (from a kitten, 21 days old) that well illustrates the division of the cerebral cortex into six horizontal layers and into vertical columns.

and their processes. It therefore is unlikely that basic differences in the organization of cell types or in the shape of dendritic trees underlie functional differences. The function of a given cortical region appears to be largely determined by its input, specifically by what sensory systems or other cortical areas make connections, and by its efferent pathways. It therefore is not surprising that changes in environmental input should have major effects on the function of the system. The input to a cortical area probably determines many of the cortical circuits that are established in that area. Whatever structural differences are seen on microscopic examination of different cortical regions appear also to be largely input related. For example, the primary visual cortex (area 17) has a large input from the eyes via the geniculocalcarine pathway. This is associated with a large and complex layer 4, the cortical layer that receives the input from the eyes via the lateral geniculate nucleus. Aside from this prominent cortical afferent system, the visual cortex looks much like cortex elsewhere. The only other remarkable feature is the absence of large pyramidal neurons in layer 5, the cortical layer that mediates the output of the cerebral cortex. This is probably related to the absence of long efferent pathways. Most of the efferent fibers of the primary visual cortex (area 17) end in neighboring visual association areas. Some descend back to the lateral geniculate nucleus. The information-processing regions of visual cortex—layers 2 and 3—have an architecture that is similar to that of other cortical areas.

The arrangement of the cerebral cortex into vertical columns or functional modules is seen especially well in the visual cortex, where alternating strips of cortex with predominant connections to one or the other eye (the ocular dominance columns) have been demonstrated (Chapter 4).

The efferent connections of a cortical region also affect the anatomy of the cortical layers to some extent. For example, layer 5 of the motor cortex is easily identifiable on the basis of the large pyramidal cells (Betz cells), which are the cells of origin of the corticospinal tract. In general, cortical regions that give rise to long efferent systems, such as the corticospinal tracts, have a relatively large fifth cortical layer containing large pyramidal cells, and a less prominent fourth cortical layer, than does cortex with large afferent inputs, such as the visual and auditory cortex. The advent of immunohistochemical staining has made it possible to identify differences in populations of neurons through the use of cell surface markers or by neurotransmitter type (Cooper and Steindler, 1986; Solodkin and Van

Hoesen, 1996). Use of this methodology has opened the way to identification of the approximate borders of other cortical regions, in addition to areas 17 (calcarine) and 4 (motor cortex).

### The Control of Dendritic and Axonal Development

In studying the development of dendrites and axons in cerebral cortex, we have to consider two events. The first is the development of neuronal shape (apical and basal dendritic trees) and the control of the orientation of neurons, which together lead to the repeating vertical modules that are characteristic of cerebral cortex. The second is the development of specific afferent and efferent systems that impart functional specificity to different cortical regions. The former is a much easier task than the latter. The author, together with B. Garber and L. Larrramendi (1980), approached the problem of development of cortical pyramidal cell shape and orientation some years ago in a tissue culture system of fetal mouse neurons that had been developed by Dr. Garber. Fetal neocortical neurons were dissociated, which resulted in a suspension of single cells without dendritic or axonal components. These cells were placed into slowly rotating flasks that contained tissue culture medium. In this system, cortical neurons aggregate into small spheres of tissue. The spheres develop an internal structure remarkably similar to that of the cerebral cortex in situ: most of the cells develop into pyramidal neurons, with apical dendrites, most of which point toward the surface of the sphere (Figure 1.5). Basal dendrites develop below the cell body, and there is an outgrowth of axons. It therefore appears that the information necessary for the outgrowth of dendrites and axons, for formation of a single main apical dendrite, and for vertical orientation and alignment of the cells is present in each fetal neuron. As a result, developing neurons will form a cortical plate with proper alignment and orientation of dendritic trees whenever a sufficient number of cells are present. This property of immature cortical neurons is defective or absent in the cortex of humans with the disease tuberous sclerosis, in which large groups of neuron precursors fail to develop orientation or polarity upon arrival at the cortical plate. These cells form the cortical malformations known as "tubers" that are characteristic of this condition (Huttenlocher and Heydemann, 1984).

The dendritic trees of cultured cortical neurons are not as extensive as are those seen in vivo, which indicates that some external factors necessary

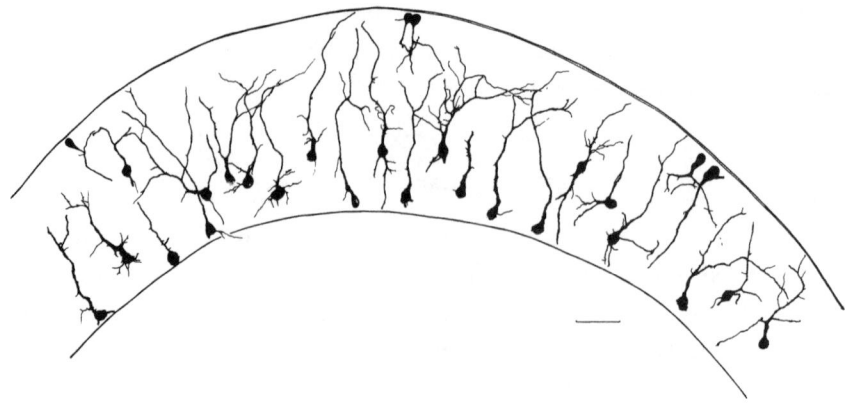

Figure 1.5. An aggregate formed in tissue culture from dissociated mouse fetal cortical neurons, shown by the Golgi method. The neurons are typical pyramidal neurons with vertical orientation and alignment, like those seen in the neocortex in situ. (From Garber, Huttenlocher, and Larramendi, 1980, fig. D, p. 267; reprinted with the permission of Elsevier Science.)

for normal dendritic growth are missing. One such factor has been identified as thyroid hormone (Honegger and Lenoir, 1980). This hormone also is of major importance for the normal growth of dendrites in vivo. In the animal or human deprived of thyroid hormone during the critical period for dendritic development (shortly before birth to at least the age of 2 years in humans), cortical dendrites fail to grow normally (Eayrs, 1971; Ipina et al., 1987). This has profound functional effects, and in the human leads to severe mental retardation. In animal models of congenital hypothyroidism, synaptic development is delayed, in addition to the defect in dendritic growth (Nicholson and Altman, 1972). Replacement therapy with thyroid hormone will prevent the abnormality in dendritic growth, if it is given during the period of normal dendritic growth. If given after this critical period, it will lead to improvement of the anatomical changes, with catch-up growth of cortical dendrites, but will not improve cortical function (Eayrs, 1971). This effect illustrates an important principle of brain development: correct timing of development is important for the establishment of normal function. The reasons for this are not well understood. One possibility is that late development of dendrites, but normal or near-normal growth of axons, will lead to a mismatching of synaptic connections because the normally arriving axons are unable to find their normal targets for innervation. Evidence of such mismatching has been reported in a study on the effects of experimental hypothyroidism on synaptic connections in the hippocampus of rats (Rami, Patel, and Rabie, 1986a, 1986b).

A major, incompletely resolved, question relates to the control of development of the specific afferent and efferent connections of the cerebral cortex, that is to say the pathfinding by developing axons. Genetic factors are of primary importance in these developmental steps. A very large number of genes appear to be involved in the signaling that leads to correct connections. It has been estimated that about a third of the 100,000 or so genes in the human genome are brain specific, and that another 20,000 or so share expression in the brain and in other organs. In other words, more than half of all our genes are related to brain development and/or function. Most of the genes that have effects in the brain appear to be important for brain development. Unraveling the functions of these genes is a daunting task, which has just begun. However, we know something about the mechanisms that must be involved from observations of developing neural systems in tissue culture and in vivo. From tissue culture experiments, similar to the one described above, we know that individual fetal neurons already contain information concerning their proper connections with other neurons, even those developing a distance away. Co-culture of neurons derived from different forebrain regions shows that these neurons segregate into groups of common origin, a process probably related to differences in cell adhesion molecules (Goodman and Shatz, 1993). In vivo experiments have shown the importance of "guideposts," cells that secrete a trophic substance toward which the growing axonal tip will orient itself. The early-developing subplate neurons appear to be such guideposts for the ingrowth of axons from the thalamus into the developing cerebral cortex (Kostovic and Rakic, 1990; Goodman and Shatz, 1993; Allendoerfer and Shatz, 1994). In general, it appears that the developing nervous system uses several cues for the guidance of axons to their proper location. These include cell surface molecules that mediate cell to substrate adhesion, molecules that inhibit cell adhesion, and diffusable trophic factors that form concentration gradients in the extracellular matrix. Trophic factors direct movement of a growing axon toward the point of highest concentration (Dodd and Jessell, 1988).

## Summary

The development of the cerebral cortex requires numerous sequential steps, which need to be precisely timed to achieve a normal outcome. The developmental steps and schedules are largely genetically determined. A huge number of genes are involved in this process. Much of modern developmental neurobiology is concerned with the unraveling of the genetics of brain development, and with the discovery of genetic defects underlying errors in brain development. Environmental factors are also important for normal development, especially the presence of a normal internal milieu, including hormones (especially thyroid hormone), growth factors, and proper nutrition. Environmental factors become more important during the later phases of development, especially those related to synaptogenesis, which will be discussed in Chapter 2.

# 2

> What mysterious forces—finally establish those protoplasmic kisses,
> the intercellular articulations, which seem to constitute the final ec-
> stasy of an epic love story?
>
> Santiago Ramon y Cajal, 1937

Synapses are the structures through which neurons communicate. The
great neuroanatomist Ramon y Cajal considered them the crowning
achievement of neurogenesis, as is clear from the quote above. Together
with axonal branches and dendrites they form the neural circuits that are
thought to underlie information processing by the cerebral cortex. Their
presence is essential for any sort of brain function. The onset of function in
the cerebral cortex therefore can be timed by the age of the formation of
the synapses.

The development and maintenance of synapses in the cerebral cortex differs from the earlier developmental steps, which are largely genetically determined. In contrast, environmental factors become important in synaptogenesis, especially in the stabilization of the initial synaptic contacts. Most
of these early synapses are thought to be made randomly, at points of contact of growing axonal branches and cortical dendrites, especially on dendritic spines (Figure 2.1A). Some of these early synapses are incorporated
into functioning circuits and are stabilized (persist), while others are useless
and are resorbed. It is here that the environmental input to the cerebral cortex becomes important for its further development. The synaptic connections that are stabilized differ depending on the environment in which a

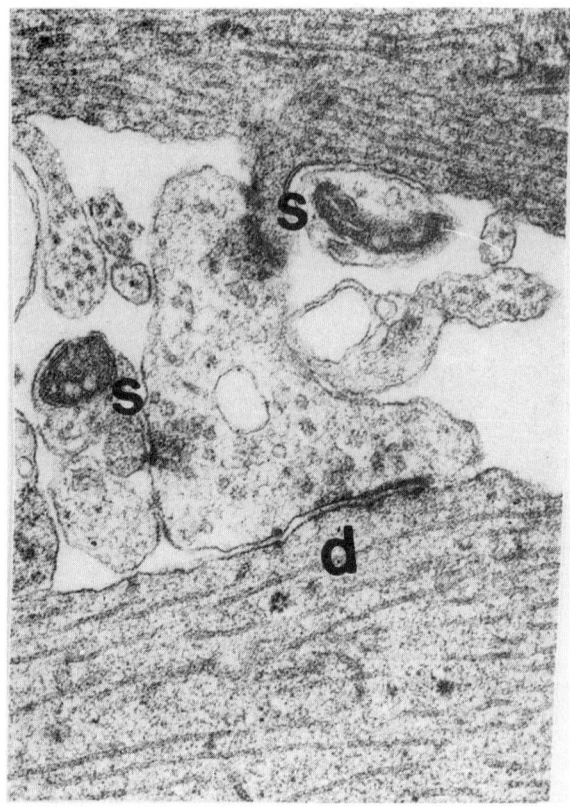

Figure 2.1A. Synapse formation in tissue culture of dissociated fetal cortical neurons after 9 days in the culture. The figure shows three early synapses, two between an axon and a dendritic spine (s) and one between an axon and the shaft of a dendrite (d). Magnification × 19,000. (From Garber, Huttenlocher, and Larramendi, 1980, fig. 5-B, p. 264; reprinted with the permission of Elsevier Science.)

child grows up. These concepts will be further developed in this and the following chapters.

### Mechanisms for Synapse Formation

The formation and stabilization of synaptic connections is a final step in the structural development of the cerebral cortex. While these processes are most rapid in infancy, they are likely to persist—at a much slower

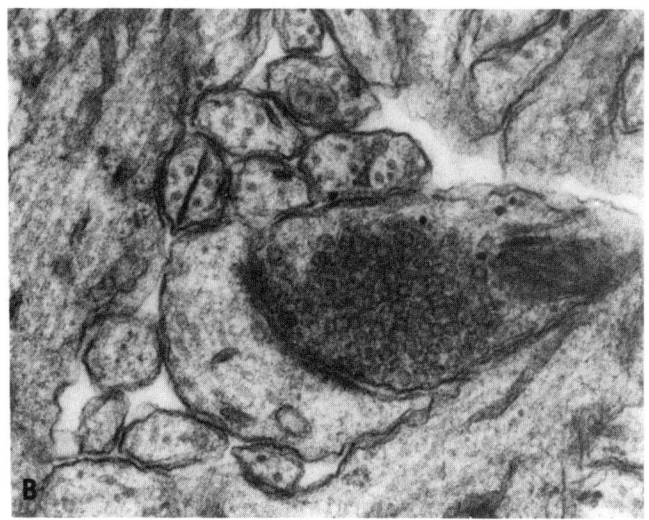

Figure 2.1B. A mature synapse in the same tissue culture as that in Figure 2.1A, demonstrating the prominent synaptic vesicles in the axon terminal and heavy staining of the specialized axodendritic membranes at the synaptic site. Magnification × 19,000. (From Garber, Huttenlocher, and Larramendi, 1980, fig. 9-B, p. 270; reprinted with the permission of Elsevier Science.)

rate—throughout the life span. Dendritic and axonal growth are closely linked with synaptogenesis. As soon as dendrites and axons grow out and touch they form synaptic connections. This is well seen in tissue culture of dissociated fetal mouse cortical neurons. These cultures initially show no axons or dendrites. Dendritic growth occurs within 2 days of placement of the cells into the culture. The first synaptic contacts appear after 4 or 5 days. (Figure 2.1). The mechanism by which this occurs was reviewed by Haydon and Drapeau (1995). A growing axon releases from its leading edge (the growth cone) a trophic substance that induces a change in the cell membrane of a nearby dendritic branch. The signaling agent often is a neurotransmitter, which later becomes localized to the presynaptic region. During synaptic development, the trophic factor induces a small patch of postsynaptic membrane specialization in the dendrite. This includes the synthesis of receptors for the neurotransmitter at the postsynaptic site, which makes it possible for the postsynaptic membrane to change in response to release of the neurotransmitter at the axon terminal, usually by a change in the resting membrane potential.

Most synaptic contacts appear to be made randomly, wherever an axon and a compatible dendrite meet. There appears also to be a second, more specific signaling system. This involves both antegrade signaling by the growing axon and retrograde signaling by the dendrite of a cell that is destined to make a specific synaptic contact ("bidirectional signaling"). This more specific mode of synaptogenesis probably is important for the development of the early neuronal circuits that mediate the most basic functions of a given cortical area, and that are relatively independent of environmental modification. In the visual cortex, some such basic circuits develop prior to birth (Hubel and Wiesel, 1963b), and therefore prior to the presence of formed retinal images. The present evidence indicates that both modes of synaptogenesis, random formation followed by functional stabilization and formation of neuronal circuits by intrinsic mechanisms such as bidirectional signaling, are important for development of the cerebral cortex. The relative contribution of each probably varies for different neural systems.

Bidirectional signaling, leading to neural circuits that form without any major influence of environmental stimulation, is likely to predominate in cortical regions that subserve basic functions, such as walking, which develop in every normal child. Initial random formation of synapses is likely to be more important in systems that subserve functions where learning and practice are important. In other words, bidirectional signaling is the likely mechanism for the laying down of neural circuits that are intrinsic or genetically determined. Randomly formed synapses, on the other hand, appear to form a substrate for the development of circuits that are dependent on environmental input. The ready availability of billions of unspecified synaptic contacts in the immature cerebral cortex may be important for the formation of the synaptic circuits underlying the development of higher cortical functions, including mathematical skills, musical ability, and language functions. It may account for the amazingly rapid, facile learning of many skills that occurs during early childhood. At this point, the utilization of labile synapses for childhood learning is only a hypothesis. Support for this hypothesis can be found in animal studies, some of which will be presented later in this chapter. These include the recent discovery of "silent synapses" in the immature brain.

Recent experimental data from the cerebral cortex of rats provide evidence for the initial production of functionally unspecified synapses. Silent synapses, that is to say synapses that do not transmit a nerve impulse under physiologic conditions, have been found in rodents' cerebral cortex between postnatal days 2 and 8. During this time window, there is increased ability of rodent cerebral cortex to mold itself in response to changes in the periphery. Up to about the postnatal age of 8 days, the number of cortical barrels (anatomically distinct regions in the sensory cortex representing the mustacial vibrissae) is adjusted to the number of whiskers (see Chapter 4). The close temporal correspondence between the presence of silent synapses and the malleability of rodent sensory cortex in response to changes in the periphery provides supporting evidence for a role of unspecified synapses in cortical plasticity (Isaac et al., 1997). Furthermore, it has been found that silent synapses may be turned into functional units by repetitive electrical stimulation. This finding underscores the importance of electrical activity for the development and maintenance of synaptic function.

### Synaptogenesis and Synapse Elimination in Human Cerebral Cortex

Anatomical studies of developing cerebral cortex provide strong evidence that exuberant synaptogenesis and synapse elimination occur in the cerebral cortex. In the human, these are postnatal developmental events. At term (40 weeks PO), most of the cortical neurons that will ever exist are in their correct location in the neocortex, but the brain is only about one quarter the size of the adult brain. Most of the connections, including dendritic and axonal branches and synapses, are as yet missing. These connections form largely during the first year of life, a period during which the size of the brain rapidly expands to near-adult size, and during which there is a tremendous burst of synapse formation, at the end of which the total number of synapses approaches twice that seen in the adult. This is followed by a much longer period of elimination or pruning of excess synaptic contacts, which is not completed until late adolescence in many cortical areas.

Some evidence for postnatal synaptic pruning has already been men-

Figure 2.2. An electron micrograph of human cerebral cortex prepared by the PTA method, showing numerous synaptic profiles against an unstained background. This is prefrontal cortex of a child 3 and 9/12 years old. Magnification × 19,000.

tioned in the discussion of the Golgi method. Electron microscopy (EM) provides a much more accurate method for the study of synapse number and density, and lends itself quite well to the generation of quantitative data. The first EM study on synaptogenesis in human cerebral cortex was published in 1973 by Molliver, Kostovic, and Van der Loos. It was concerned with the onset of synaptogenesis in the neocortex. Very well preserved fetal tissue was studied with conventional EM techniques. It was found that synapse formation in human prefrontal cortex begins in the second trimester, prior to PO week 20, that is to say prior to the completion of neuronal migration. The first synapses are found not in the cortical plate, but in the molecular layer, just below the pia (covering membrane) of the cerebral cortex and in the subplate zone (the white matter just below the cortex) (Kostovic and Molliver, 1974). Kostovic, Seress, and their colleagues (1989) subsequently showed an even earlier onset of synaptogenesis in the hippocampus, in fetuses at about 15–16.5 weeks, mostly at the upper limit of the cortical plate.

Data on later phases of synaptogenesis were first published in 1979 (P. R. Huttenlocher, 1979). This study and most subsequent ones on human synaptogenesis utilized the phosphotungstic acid (PTA) method (Bloom and Aghajanian, 1968), which stains perisynaptic proteins selectively. This results in the demonstration of "synaptic profiles" that are easily recognizable against a nearly unstained background (Figure 2.2). A profile consists of two or more presynaptic projections, a thin intracleft line, and a postsynaptic band. Approximate synaptic density values can be calculated from synaptic profile counts, by use of a correction factor which corrects for the fact that some profiles may extend over more than one section and would therefore be counted twice, while others are cut at angles that do not yield recognizable structures. The PTA method is preferable to conventional EM in human brain tissue obtained at autopsy because the perisynaptic proteins, which it demonstrates, are very resistant to postmortem autolysis. Synapses are difficult to count in conventional EM prints from autopsy tissue, in view of the autolytic changes in other synaptic components, such as synaptic vesicles. In perfusion-fixed animal brain tissue the PTA method yields synaptic density values that are similar to those obtained by conventional EM (Aghajanian and Bloom, 1967; Bloom and Aghajanian, 1968; Armstrong-James and Johnson, 1970).

## THE PREFRONTAL CORTEX

The initial study utilizing the PTA method was carried out in layer 3 of the middle frontal gyrus, in the prefrontal cortex. It showed what was at that time a surprising finding: synaptic density was much higher during late infancy and childhood than in the adult years. During early infancy, there was a rapid rise in synaptic density, which approached a maximum at about the age of 1 year. This was followed by several years during which synaptic density stayed high, with decrease to the adult range some time between the age of 7 years and mid-adolescence (Figure 2.3). The neuronal density was determined in the same cortical region as the synapse quantitation, that is to say in layer 3 of the middle frontal gyrus. The neuronal density was found to decrease rapidly during the period of initial synaptogenesis, which is in the first postnatal year. This decrease in the density of neurons is at least in part due to the expansion of the neuropil (dendrites, axonal branches, glial fibers, and synapses), which results in a significant increase in the total cortical volume. The availability of neuronal and synaptic density data made it possible to calculate the approximate number of synapses per neuron in layer 3. The number increased rapidly during the first postnatal year, reaching a maximum of about $10 \times 10^4$ synapses per neuron early in the second year of life, followed by a decline to about $8 \times 10^4$ in the young adult. The results suggested pruning of synapses from individual neurons. More extensive data related to synaptic pruning have been obtained more recently in studies of the visual cortex (see below).

The number of brains examined was small, especially in the childhood years, because the method requires samples of unfixed brain tissue, as well as tissue that is relatively free of antemortem and postmortem changes, which rarely becomes available in childhood. Another problem was the absence of information about the total volume of brain tissue in which the synaptic density was measured. A decrease in synaptic density during childhood could be due either to synapse elimination or to expansion of the cerebral cortex, or both.

## THE VISUAL CORTEX (AREA 17)

As pointed out above, it has, until recently, not been possible to define the exact boundaries of most cortical areas. The visual cortex (area 17) is an exception, since it is defined by the presence of a large fourth cortical layer, bi-

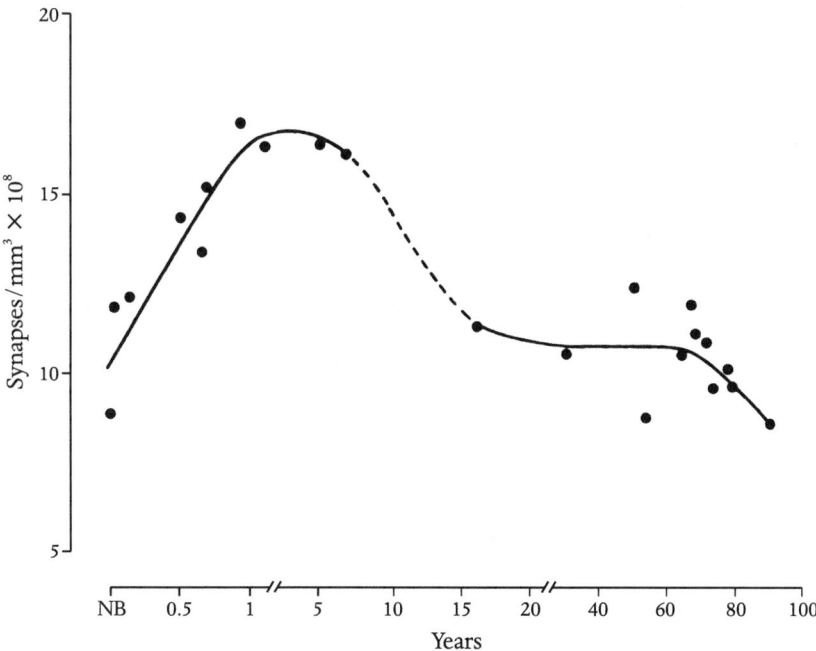

Figure 2.3. Synaptic density in human prefrontal cortex (middle frontal gyrus) as a function of age. The maximum synaptic density in this cortical area is reached at about the age of 1 year. The dashed line indicates insufficient data. (From P. R. Huttenlocher, 1979, fig. 3; reprinted with the permission of Elsevier Science.)

sected by a bundle of axons, the line of Genari. In the visual cortex, it therefore is possible to determine cortical volume changes related to age. In collaboration with Drs. Christian deCourten, Hendrik Van der Loos, and Lawrence Garey at the Institute of Anatomy in Lausanne, Switzerland, my colleagues and I carried out a study on age-related changes in the synaptic density and volume of the visual cortex, using computer-assisted methods for synapse quantitation and for area measurements (Huttenlocher, deCourten, Garey, and Van der Loos, 1982; Huttenlocher and deCourten, 1987). The synaptic density showed age-related changes similar to those in the prefrontal cortex: an initial rapid increase in synaptic density in early infancy, followed by a period of high synaptic density, followed by a decrease to adult values, which were about 60 percent of the maximum. However, the timing of these changes differed from those in the prefrontal cortex. In the visual cortex, there was a sudden burst of synaptogenesis between the

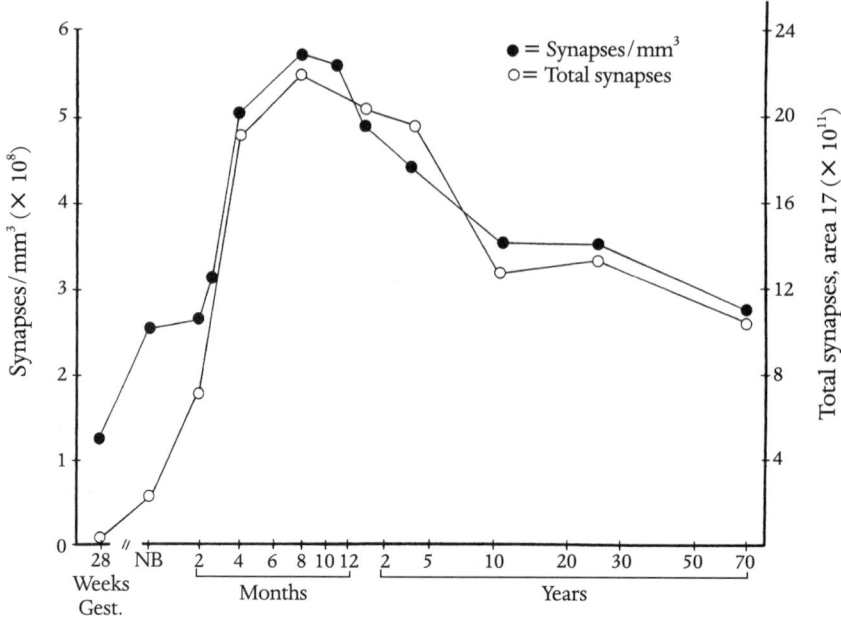

Figure 2.4. The synaptic density and the total number of synapses in human visual cortex (area 17) as a function of age. (From P. R. Huttenlocher, 1990, fig. 1, p. 519; reprinted with the permission of Elsevier Science.)

postnatal age of 2–4 months. The subsequent decline in synaptic density occurred earlier than in the prefrontal cortex, between the ages of 12 months and 10 years (Figure 2.4).

The volume of area 17 was found to increase at the time of rapid synaptogenesis, followed by a slight decline during late childhood (Figure 2.5). A decline of the visual cortex volume during childhood has also been reported by Sauer and colleagues (1983), who used measurements from postmortem human brain tissue. Jernigan and associates (1991), using quantitation from magnetic resonance imaging (MRI) scans, found a developmental decrease in the prefrontal cortical volume with this noninvasive in vivo method, which can be used for the study of normal developmental events in the human. This decrease in volume occurred during adolescence, somewhat later than that in the visual cortex, consistent with the late occurrence of developmental events in human prefrontal cortex. Pfefferbaum and colleagues (1994) found a decrease in total gray matter volume starting at about the age of 10 years. These findings were confirmed by Rajapakse,

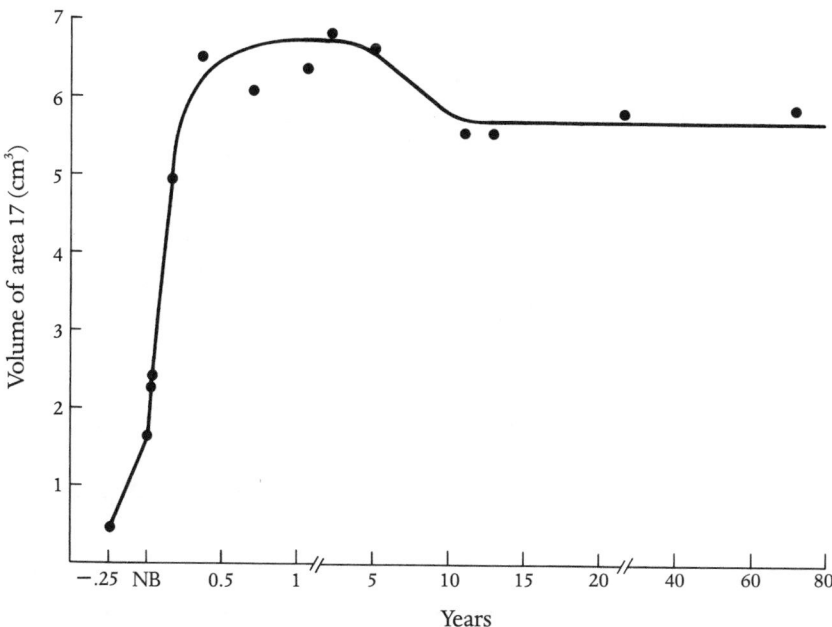

Figure 2.5. The volume of the visual cortex as a function of age. The cortical volume between ages 5 months to 5 years is significantly above the adult value ($p < .05$).

DeCarli, McLaughlin, and their colleagues (1996), who found a decrease in gray matter volume of human cerebral cortex between the ages of 4 and 18 years, while both the volume of white matter and the volume of fluid-filled spaces showed increases. The timing of in vivo quantitative MRI data on the shrinkage of the cerebral cortex correlates well with that of postmortem anatomical findings on synapse elimination. It therefore appears that the MRI method may be useful for the noninvasive estimation of the time course of synaptic and dendritic pruning in different regions of developing human cerebral cortex.

From our synaptic density and volume data on the visual cortex we were able to calculate the total number of synapses in area 17. This increased from about $2 \times 10^{11}$ at term to a maximum of about $22 \times 10^{11}$ at the age of 6 months, that is by a factor of 10. A remarkable burst of synapse formation occurs in human primary visual cortex during this period; on the order of $10^{10}$ new synapses are formed per day, or 100,000 per second! The total number of synapses in primary visual cortex subsequently decreases to a value of about $14 \times 10^{11}$, a value that is maintained during the adult years (Figure 2.4). This study proved that the observed decrease in synaptic density during childhood is due to loss of synapses rather than to expansion of neuropil, since shrinkage rather than expansion of cortical volume occurs at the time of the decrease in synaptic density. It is therefore well established that the immature human cerebral cortex has a great excess of synaptic connections over those remaining in the adult. It is likely that initial overproduction of synapses followed by synapse elimination occurs to some extent in all developing neural structures (Purves and Lichtman, 1980). The possible functional importance of this difference between the immature and the adult brain will be discussed later.

### Synaptic Pruning

Leuba and Garey (1987) obtained neuronal density values in all layers of the visual cortex from the same brains studied by us. The resulting data on synapses per neuron show significant pruning of synapses from individual cortical neurons, supplementing the more limited data on the middle frontal gyrus. This is seen in all cortical layers to approximately the same extent. The mean number of synapses per neuron varies in different cortical layers, and it is related to variations in neuronal size, which is least in layer 4, where small stellate neurons predominate. Figure 2.6 shows the number of synapses per neuron as a function of age in different layers of the primary visual cortex. This number rises very rapidly between the ages of 2 and 8 months, when it reaches a maximum, which is nearly 20,000 in layer 3, nearly 7,000 in layer 4c. Subsequently, there is a decline in the number of synapses per neuron, related to synaptic pruning. In the prefrontal cortex, data are available only for layer 3 (Huttenlocher, 1979). Neurons in layer 3 of the prefrontal cortex have larger numbers of synapses per neuron, and this number is reached more slowly than the maximum in the calcarine cortex. The maximum number of synapses per neuron in human prefrontal cortex is about 100,000, five times the number in the calcarine cortex, and this maximum number is reached during the second postnatal year. In general, the number of synapses per neuron appears to be greatest in human association cortex, and fewer in primary sensory areas such as the calcarine cortex. Even in human calcarine cortex, this number is about twice that found in the visual cortex of the cat (Winfield, 1981).

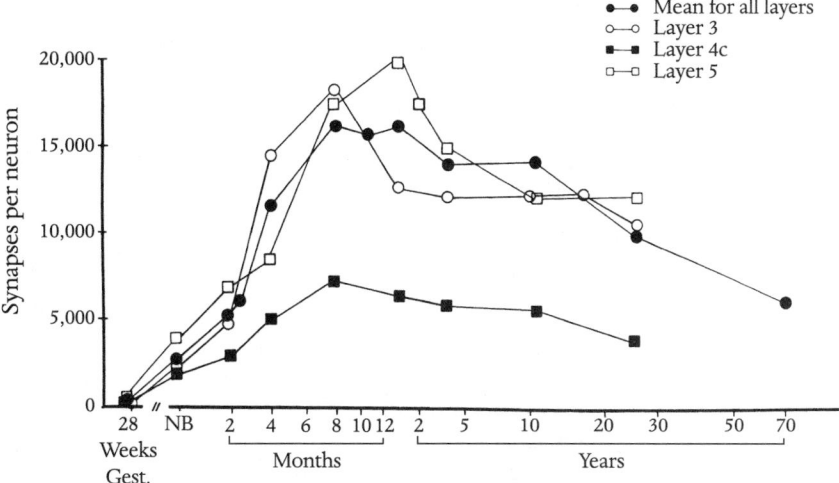

Figure 2.6. The mean number of synapses per neuron in different layers of human visual cortex (area 17) as a function of age. (From P. R. Huttenlocher, 1990, fig. 2, p. 521; reprinted with the permission of Elsevier Science.)

Synaptic pruning may eliminate whole types of input to cortical neurons. For example, the Purkinje cells of the cerebellum initially have synapses on the cell body (axosomatic synapses) as well as axodendritic synapses. These direct connections between afferent fibers and the cell body disappear during development. In the mature Purkinje cell, depolarization of the cell body and firing of the cell therefore are dependent on input via axodendritic synapses (Larramendi, 1969).

### Synaptogenesis in Other Cortical Areas

Recently, I have studied synaptogenesis in several other areas of human cerebral cortex (P. R. Huttenlocher, 2000). The results from studies of the language areas have been especially interesting. The development of synapses in the auditory cortex (Heschl's gyrus) precedes that in the receptive language areas (Wernicke's area), which in turn precedes synaptogenesis in the motor speech (Broca's) area (Figure 2.7). Synaptogenesis appears to reflect the sequence of functional development in these regions of the cerebral cortex, since development of the response to sounds precedes language comprehension, which in turn precedes the development of motor speech.

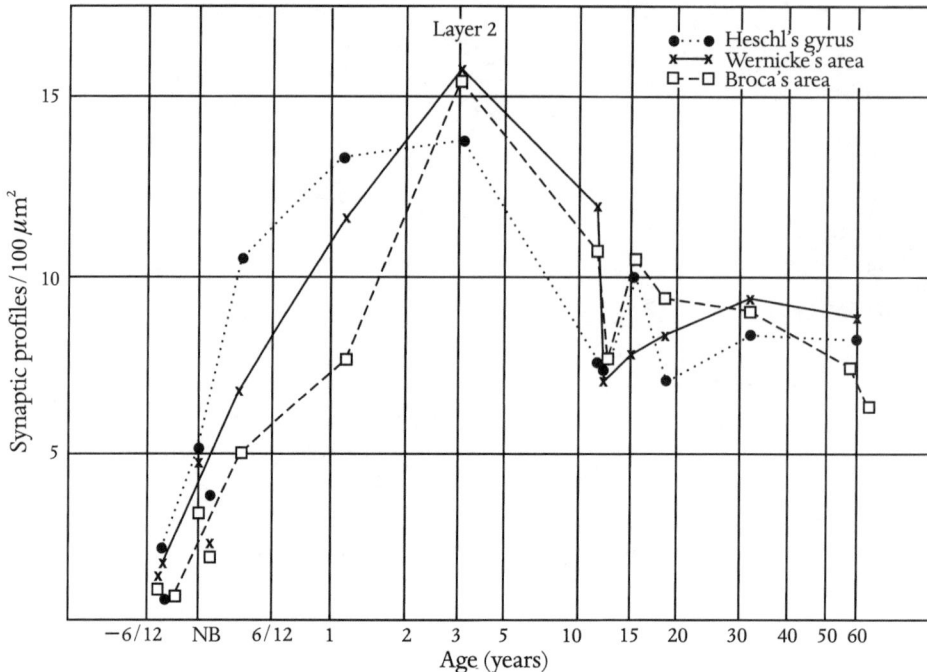

Figure 2.7. Synaptic density as a function of age in three cortical regions important for language processing. Dotted line = the auditory cortex (Heschl's gyrus); solid line = Wernicke's area in the left temporal lobe; dashed line = Broca's motor speech area in the left frontal lobe.

These functions appear to emerge during the phase of rapid increase in synaptic density.

The motor cortex (hand area) is another region in which synaptogenesis has been studied recently. It differs from other neocortical areas in that synaptic density is lower, especially in layer 5, which contains the large Betz cells. There also appears to be little in the way of synaptic pruning in the lower layers of the motor cortex. This difference from other cortical regions, and possible effects on function and plasticity, are discussed in Chapter 5.

### Apoptosis in Human Cerebral Cortex

Apoptosis is the normal or programmed death of cells during development. It removes unnecessary or redundant cell populations. In developing ner-

*Table 2.1.* Volume, neuronal density, and total number of neurons in the striate cortex (area 17) at various ages. Reprinted from P. R. Huttenlocher, "Morphometric study of human cerebral cortex development," *Neuropsychologia*, 28 (6): 517–527, copyright 1990, with the permission of Elsevier Science.)

| Age | Volume (mm$^3$) | Neurons/mm$^3$ ($\times 10^4$) | Total neurons ($\times 10^7$) |
|---|---|---|---|
| 28 weeks GA | 230 | 61.9 | 14.2 |
| 40 weeks GA | 920 | 9.6 | 8.8 |
| 6 days | 1,270 | 9.4 | 11.9 |
| 2 weeks | 1,330 | 10.9 | 14.5 |
| 2 months | 2,720 | 5.2 | 14.1 |
| 4 months | 3,800 | 3.5 | 13.5 |
| 19 months | 4,170 | 3.0 | 12.5 |
| 3 9/12 years | 4,430 | 3.1 | 13.7 |
| 5 years | 4,310 | 2.5 | 10.8 |
| 11 years | 3,620 | 2.5 | 9.1 |
| 13 years | 3,620 | 2.9 | 10.5 |
| 26 years | 3,790 | 3.6 | 14.6 |
| 71 years | 3,850 | 4.2 | 16.2 |

vous systems, apoptosis removes neurons that fail to find targets for inner-vation on other neurons. Curiously, data on the visual cortex, derived from studies of a series of brains in which both the neuronal density (Leuba and Garey, 1987) and the volume (Huttenlocher and DeCourten, 1987; Hutten-locher, 1990) of area 17 had been determined, showed no evidence of "pro-grammed neuronal death" or apoptosis in the age range studied, specifically between the conceptual age of 27 weeks and maturity (Table 2.1).

More recent measurements of visual cortex volume and neuronal density in 46 human brains between mid-gestation and old age also failed to show neuronal loss (Leuba and Kraftsik, 1994). There was marked variability in neuronal number in human visual cortex, unrelated to age, with a normal range of about 8 to 20 $\times 10^7$ neurons. This large normal variability in hu-man visual cortex would make it difficult to detect small age-related de-creases in neuronal number. A study by Klekamp and colleagues (1991) comes to a very different conclusion. These authors report a 43 percent reduction in neuronal numbers postnatally. However, a semiautomatic neu-ronal counting method was used, which did not provide reliable results for infant brains under the postnatal age of 2 months, owing to the failure to reliably distinguish small neurons from glia cells. Such an error would be expected to result in spuriously high neuronal counts in the infant brains. This method may also have undercounted neurons in mature brains, where a mean of about $11 \times 10^7$ visual cortex neurons were found by the semiautomatic counting method, versus about $14 \times 10^7$ by manual counting (Leuba and Kraftsik, 1994). A method that overestimates the number of immature neurons and underestimates the number of more mature neurons could generate data that mimic the expected data if there was apoptosis during childhood. The question of the extent and timing of apoptosis in the human cerebral cortex has to be considered unsettled at present.

In the mammalian nervous system, apoptosis has been documented best in subcortical neuronal populations. A good example is the neuromuscular system, consisting of anterior horn cells, neuromuscular synapses, and skeletal muscle. In the development of this system, axons of anterior horn (motor) neurons must find a specific target for innervation (a muscle cell or myocyte). Target finding is crucial in such a program, and neurons that fail to find a specific target undergo apoptosis. The nervous system that results from such a program would be expected to be highly efficient for a specific function (such as neuromuscular transmission), but to have limited flexibility. Variations in the development of such a simple nervous system would depend to a large extent on the number of available targets for innervation. This appears to be the case in the spinal neuromuscular system. In the mature state, each muscle cell has innervation from only a single spinal motor neuron, although each spinal motor neuron innervates a large number of muscle cells (about 100 in limb muscles). Initially, several motor neurons may make synaptic contact with a single muscle cell. However, the excess synapses are withdrawn, until only innervation from a single neuron remains (Brown, Jansen, and Van Essen, 1976). Spinal motor neurons that do not find muscle cells as targets for innervation undergo apoptosis (Table 2.2). The removal of targets for innervation by amputation of a limb increases apoptosis and decreases the population of anterior horn cells that survive to maturity, while the grafting of an extra limb prior to the normal occurrence of apoptosis increases the number of anterior horn cells that survive to maturity (Hamburger, 1975; Hollyday and Hamburger, 1976).

In complex nervous systems, such as human neocortex, available targets appear to be plentiful. The removal of one target for innervation leads not to apoptosis, but to strengthening of synapses on other targets (Bhide and

*Table 2.2.*  Programmed cell death (apoptosis).

| Cell/animal | Percent loss | References |
|---|---|---|
| Spinal motoneurons (frog) | 75 | Hughes, 1961 |
| Spinal motoneurons (chick) | 40–50 | Hamburger, 1975 |
| Trochlear nucleus (chick) | 50 | Cowan and Wenger, 1967 |
| Retinal ganglion cells (monkey) | 50 | Rakic and Riley, 1983 |
| Retinal ganglion cells (human) | 50 | Provis et al., 1985 |
| Cerebral cortex (mouse) | 30 | Heumann and Leuba, 1983 |
| Cerebral cortex (monkey) | 15 | O'Kusky and Colonnier, 1982 |
| Visual cortex (human) | None found | P. R. Huttenlocher, 1990; Leuba and Kraftsik, 1994 |

Frost, 1992). For example, Innocenti (Innocenti 1981, 1995; Innocenti, Fiore, and Caminiti, 1977) found that neurons in the occipital cortex that send transient axons to the opposite hemisphere do not undergo apoptosis after these axons are withdrawn. Instead, these neurons appear to make ipsilateral connections (O'Leary, Stanfield, and Cowan, 1981; O'Leary, 1992). A similar reaction—target finding on other neuronal populations—is seen in neurons that form transient corticospinal connections originating in the occipital cortex of rats (Stanfield and O'Leary, 1985b).

Animal studies shed some light on the question of whether apoptosis is a significant developmental event in the cerebral cortex. A small postnatal neuronal loss, about 16 percent of the total, was reported by O'Kusky and Colonnier (1982) in the visual cortex of the monkey. A more substantial loss of about 30 percent was found in the cerebral cortex of the mouse (Heumann and Leuba, 1983).

In general, it appears that apoptosis is a more important factor in simple systems, such as the spinal cord motor neurons (anterior horn cells), where about 50 percent of the neuronal population is wiped out by apoptosis (Hamburger, 1975; Hollyday and Hamburger, 1976), than in primate cerebral cortex, where apoptosis appears to occur in less than 20 percent of neurons. Apoptosis in the cerebral cortex appears to be more of a factor in mammals with simple brains, specifically rodents, than in primates (Heumann and Leuba, 1983; Leuba and Garey, 1987; Leuba and Kraftsik, 1994) (see Table 2.2). The low numbers of neurons undergoing apoptosis in primate cerebral cortex may relate to the great complexity of this structure, with availability of numerous targets for innervation (specifically dendrites of other cortical neurons and of subcortical nuclei) which may make target finding a relatively trivial task.

Programmed cell death does appear to occur in the human during fetal life in specific populations of neurons near the cerebral cortex, specifically in the subplate neurons, located in the white matter just below the cortical plate and in the Cajal-Retzius cells in the molecular layer. These two cell populations, which are very evident in the embryonic human brain, are reduced in number in the mature cerebral cortex. These are small populations of neurons, whose loss would not be evident in neuronal counts of the entire cerebral cortex. In the embryo and fetus, they are located in regions devoid of other neurons, and they may therefore be unable to find sufficient targets for the innervation needed for survival.

The recent discovery of new neuron formation in adult primate neocortex (Gould et al., 1999) makes it necessary to rethink the question whether counts of the neuronal number during infancy, childhood, and the adult years provide an adequate index of neuronal loss, because formation of new neurons may occur at the same time that other neurons disappear. An absence of change in neuron number therefore may be due either to the absence of apoptosis or to apoptosis, approximately balanced by the formation of new neurons.

Data obtained from studies of the embryonic rodent brain suggest that apoptosis in the cerebral cortex may occur largely during early embryogenesis, prior to the developmental stages studied by Leuba and Garey (1987) and Leuba and Kraftsik (1994) in humans. Kuida and colleagues (1998) bred genetically engineered mice ("knock-out mice") that lacked caspase 9, an important component of the mammalian programmed cell death machinery. These mutant mice had overgrowth of the cerebral cortex to the point that the cortex showed formation of convolutions with gyri and sulci, where normally the rodent brain is smooth (lissencephalic). The overgrowth of the cerebrum was seen already at E. (embryonic day) 10.5, at an early developmental stage, which suggests that apoptosis normally occurs very early in the formation of the cerebral cortex. The animals with overgrowth of the brain did not survive past the neonatal period, which indicates that in the brain more is not necessarily better. Abnormal enlargement of the brain in humans (megalencephaly) is also associated with reduced cortical function, manifested as mental retardation.

*Table 2.3.* Synapse elimination.

| Cortex/animal | Percent loss | References |
|---|---|---|
| Neocortex (rat) | <10 | Aghajanian and Bloom, 1967 |
| Visual cortex (cat) | 33 | Cragg, 1975 |
| Visual cortex (monkey) | 40–50 | Rakic et al., 1986 |
| Visual cortex (human) | 42 | P. R. Huttenlocher et al., 1982 |

### Synaptogenesis in Human Cerebral Cortex versus That in Other Species

It is of interest that, contrary to programmed neuronal death, synaptic pruning and loss of synapses during postnatal brain development appear to occur to the greatest extent in the most complex neural systems (Huttenlocher, 1994). Developmental loss of synapses is highest in human and monkey cerebral cortex, less in the cortex of the kitten (Cragg, 1975b), and still less in the immature rat brain (Bloom and Aghajanian, 1967) (see Table 2.3). These differences may be secondary to the formation of more random, unspecified early synapses in the more complex brains.

In simple systems, connections may be made to a large extent through genetically determined programs, which do not require large numbers of unspecified synapses.

The development of higher mammalian neocortex is likely to be to a large extent shaped by input to the system, which is to say that it has a high degree of functional plasticity (O'Leary, 1989). Such a developmental program would be predicted to leave large numbers of functionally unspecified synapses that may disappear. There may, therefore, be significant differences between humans and other mammals in the way in which functional neuronal circuits in the cerebral cortex are established. Some differences are apparent even when one compares synaptogenesis in human and in other primate cerebral cortex.

### Synapse Formation and Elimination in the Cerebral Cortex of the Rhesus Monkey

Very extensive data on synapse formation in the cerebral cortex of the rhesus monkey have been published by P. Rakic, P. Goldman-Rakic, P. Bourgeois and their colleagues at Yale (Rakic et al., 1986; Bourgeois and Rakic, 1993; Rakic and Bourgeois, 1993; Bourgeois, Goldman-Rakic, and Rakic, 1994; Rakic, Bourgeois, and Goldman-Rakic, 1994a). The data from rhesus monkeys are based on synapse counts from conventional EM prints, obtained serially throughout the depth of the cerebral cortex. The use of conventional EM made it possible to obtain data on types of synapses (symmetric synapses, which are thought to be largely inhibitory, versus asymmetric, mainly excitatory ones). Furthermore, the proportion of cerebral cortical volume that is occupied by neuropil (axons, dendrites, synapses, and glial fibers) versus neuronal and glial cell bodies and blood vessels could be assessed. These components of the cerebral cortex cannot be accurately identified in tissue prepared by the PTA method used in studies of humans. The primate data contain synaptic density values for neuropil as well as for the cortex as a whole, while the data from the studies of humans are density values for the whole cortex only.

Comparisons of the human and monkey data show several similarities: (1) both show a phase of rapid synaptogenesis in early infancy; (2) this is followed by a plateau, during which synaptic density is above adult levels, that extends to late childhood; (3) there is synapse elimination during late childhood or adolescence and (4) the magnitude of synapse overproduction is similar in humans and monkeys. There are also two major differences: First, the time course of synaptogenesis and synapse elimination is much longer in the human than in the rhesus monkey. This appears to be the case especially in systems subserving so-called higher cortical functions, such as the prefrontal cortex, where synaptogenesis in the human extends over more than 1 year, while synapse elimination is not complete until about the age of 16 years, versus synaptogenesis occurring over about 4 months and synapse elimination over about 12 months in the rhesus monkey. Second, in the human, there are region-specific differences in the time course of synaptogenesis and of synapse elimination (Huttenlocher and Dabholkar, 1997; Huttenlocher, 1990, 1994) (see Figures 2.7 and 2.8). In general, the primary sensory cortex (visual, auditory) develops earlier than the motor and association cortex. In the classic language areas, receptive language areas (Wernicke's area in the left posterior temporal cortex) develop somewhat earlier than the motor speech (Broca's) area in the left prefrontal cortex (P. R. Huttenlocher, 1999). No such region-specific differences were seen in

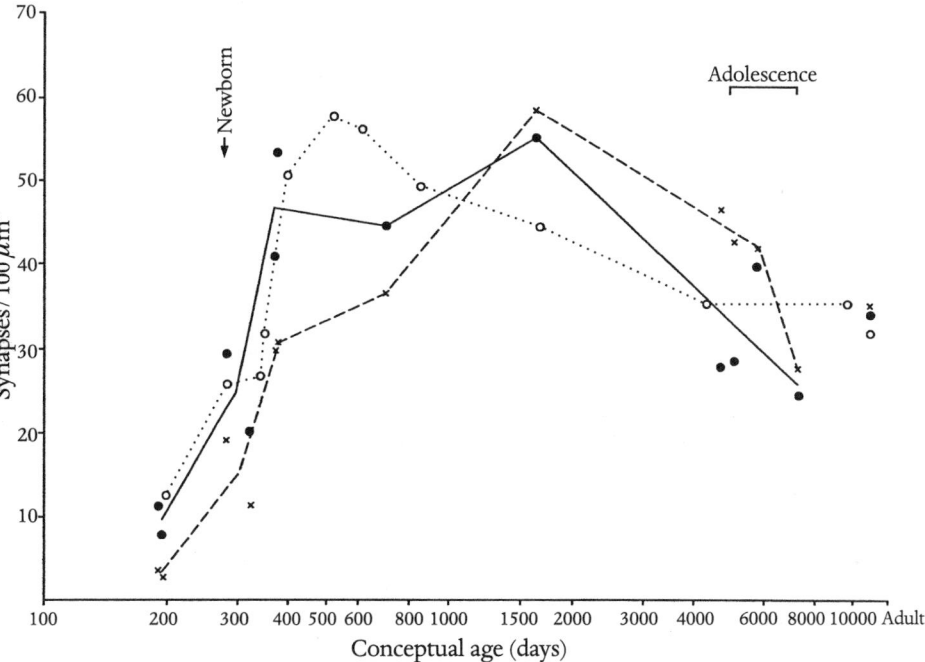

Figure 2.8. The Hierarchical pattern of synaptogenesis in human cerebral cortex. Synapse formation in visual (open circles) and auditory cortex (filled circles) precedes that in the prefrontal cortex (middle frontal gyrus, the crossbars). (From P. R. Huttenlocher and Dabholkar, 1997, fig. 2, p. 170; reprinted with the permission of John Wiley & Sons, Inc.)

the rhesus monkey, where synaptogenesis appears to be concurrent in all cortical areas.

### A Connectionist Model for Synaptic Development

A considerable body of evidence supports a "connectionist" model for the development of a functional system from a system with initially mainly random connections. The strength of the connections between units (neurons) in a complex neural net, containing both excitatory and inhibitory elements, appears to be shaped by the input to the system. Computer simulations have shown that a randomly connected nerve net may develop organized activity and a predictable output, if it is provided with ordered (complex but not random) inputs (see Chapter 3). The available evidence suggests that the development of functioning neuronal circuits follows a similar pattern. As has been pointed out, initial synaptogenesis in the cerebral cortex is not under environmental control. Premature delivery of the fetus, providing earlier exposure to complex visual stimuli, does not change the time course of synaptogenesis in the visual cortex of the monkey (Bourgeois, Jastreboff, and Rakic, 1989). The analogy to computer simulation would be that neurons, axons, dendrites, and synapses that are formed early correspond to the state of the computer prior to programming, when excitatory and inhibitory units are randomly connected. In contrast, the subsequent developmental steps, namely the development of specific neural circuits and the pruning of excess (unused) synapses, are thought to be largely dependent on input from the environment. They correspond to changes in the connection strength between units in the computer, and are related to recurrent external input.

### Animal Studies That Show Effects of the Environment on Synapse Formation and Elimination

The classic studies by David Hubel and Torsten Wiesel on the effects of visual deprivation in the infant cat and monkey provide an excellent example of the environmental control of synaptogenesis and synapse elimination. In the developing visual cortex, the specific geniculo-cortical connections that are made and/or maintained are dependent on visual input. When one eye is deprived of formed visual images, most of the neural connections in the cerebral cortex are made with the seeing eye (Chapter 4).

A study by Winfield (1981) provides interesting data on the effects of postnatal visual deprivation on synaptic density in the visual cortex. In this study, synaptic density in kittens deprived of formed visual images by unilateral and bilateral eyelid suture was compared to that in control kittens. In the control animals, synaptic density increased rapidly from birth to the age of 40 days. The maximum synaptic density, about $3.8 \times 10^8$ synapses/millimeters$^3$, was reached by the age of 70 days. This was followed by synapse elimination, with a decrease in density to $3.2 \times 10^8$ synapses/millimeters$^3$ by the age of 110 days, and $2.5 \times 10^8$/millimeters$^3$ in the adult. The results seen in the bilaterally visually deprived animals were especially interesting. Early synaptogenesis paralleled that in the controls. At the age of 40

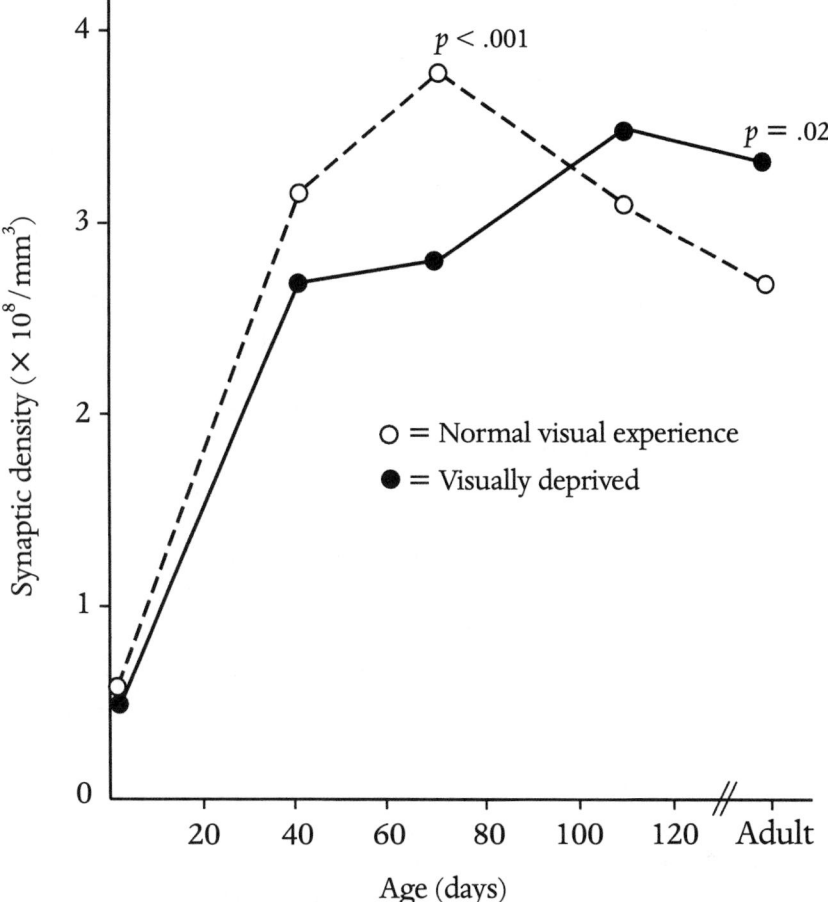

Figure 2.9. The effects of postnatal deprivation of visual input by eyelid suture on synaptic density in the visual cortex of the kitten. Solid line = control animals; dashed line = visually deprived animals. (Data from Winfield, 1981.)

days, synaptic density in the visually deprived animals was only slightly (15 percent) below normal (Figure 2.9). This agrees well with results obtained by Cragg (1975c), who found synaptic density values about 10 percent below normal in lid-sutured animals at the age of 45 days. Subsequently, the rate of synaptogenesis slowed, and by the age of 70 days synaptic density in the visually deprived animals was 26 percent below normal. More striking, there was little evidence of synaptic pruning in the visually deprived animals. Instead, synaptic density continued to show a slow increase, reaching a value 13 percent *above* normal by the age of 110 days, and 22 percent above normal in the adult animals (Figure 2.9). This study provides evidence that synaptic pruning is indeed influenced by environmental input. The direction of the observed effect is surprising. One might have expected increased synapse elimination in the cortex that has lost its input. However, we now know that immature visual cortex, deprived of its normal input from the eyes, develops other afferent connections, including somatosensory ones, which form functioning systems. As a result, early visual deprivation leads to a high level of metabolic and hence probably synaptic activity in the primary visual cortex (Chapter 4).

Earlier in development, during the fetal period, environmental influences are fewer, but are not completely absent. Removal of the eyes in fetal monkeys results in early exclusion of all retinal input to the visual cortex. This leads to several cortical malformations, including reduction in the size of the primary visual cortex and failure of the normal developmental loss of callosal connections in secondary visual areas (Dehay et al., 1989). Even in this extreme case of removal of the eyes in the fetus, it is a late developmental event, specifically the pruning of normally transient callosal connections, that appears most strikingly affected. These findings support the conclusion that early phases of synaptogenesis are relatively independent of environmental input, while later phases, and especially synaptic pruning, are under environmental control. A more recent study of two monkeys with removal of the eyes in the embryo (at 59 and 67 days of gestational age) showed a small primary visual cortex, as had been reported by Dehay and colleagues, but normal distribution and density of the synapses at the ages of 3 months and 3 years (Bourgeois and Rakic, 1996). The only exception was layer 4b and 4c, where the ratio of synapses on the dendritic shafts versus dendritic spines was reversed, resembling the normal immature pattern. In other words, the visual cortex, disconnected from the eyes during its entire developmental period, acquired normal synaptic density and organization except in layers that normally receive geniculo-cortical input. This input is greatly diminished after removal of the eyes. The occurrence of normal synaptogenesis in the remaining visual cortex may be related to the specification of this cortical area for other sensory functions in the congenitally blind (Chapter 4).

Earlier development of the primary sensory cortex would be predicted if cortical development follows a hierarchical pattern. In such a developmental program, the formation of functional circuits in the primary sensory areas would depend to a significant extent on sensory input. The development of secondary sensory and association areas would depend on input from the primary sensory areas, and could not proceed until this input became functional. Finally, the prefrontal cortex, which mediates the integration of information from multiple association areas, would develop functional systems only after some input from these association areas had been established. Hierarchical development of synaptic connections in human cerebral cortex supports the hypothesis that synaptogenesis and synapse elimination are related to function and depend on input, either from sensory systems or from other brain areas. This carries the implication that the environment can influence cortical development.

Concurrent synaptogenesis, on the other hand, carries the implication that synapse formation and synapse elimination are largely independent of input to the system and occur either by genetically determined programs or on a random basis. The early phase of synapse formation probably is genetically determined and/or occurs randomly both in humans and in lower primates. This is suggested by the onset of synaptogenesis, which occurs early in fetal life in all cortical regions tested, including the visual cortex, at a time when external visual input is as yet absent. In the human, synaptogenesis proceeds at about the same rate in the visual and prefrontal cortex until shortly after birth, when synaptogenesis in the visual cortex accelerates, with the maximum number of synapses reached by about the age of 5 months. Synapse formation in the prefrontal cortex is slower, with the maximum reached much later, after the age of 1 year. The divergence in development of the two cortical areas increases after the age of 1 year, with synapse elimination in the visual cortex beginning already at that age, while it is delayed until late childhood in the prefrontal cortex. Heterochronous development of the cerebral cortex in humans is supported by data on cortical volume, which show earlier growth of the primary visual cortex than of the cortex as a whole and of the hippocampus (Sauer, Kammradt, Krauthausen, et al., 1983).

## The Relationship between Synaptogenesis and Myelination of the Subcortical White Matter

In the human, anatomical evidence for a hierarchical system of development is seen in structures other than synapses, including dendritic development as demonstrated by the Golgi methods, presented in Chapter 1. A similar hierarchical pattern exists in the myelination of the subcortical white matter.

Extensive data on myelination of the subcortical white matter in the human are contained in the Yakovlev collection of human whole brain sections at the Armed Forces Institute of Pathology (AFIP) (Yakovlev and LeCours, 1967). These show early myelination, present already at term, in fiber tracts related to the visual and somatosensory cortex. The prefrontal white matter myelinates last and very slowly. Semi-quantitative data, which include the normal range of myelination at a given age at multiple anatomical sites, have been published by Brody, Kinney, Kloman, and Gilles (1987) and by Kinney, Brody, Kloman, and Gilles (1988). Recently, histologic data have been supplemented by those from MRI, a noninvasive method that provides an accurate representation of cerebral anatomy. Several monographs on myelination as studied by MRI are available (Salamon, 1990; Wolpar and Barnes, 1992). These show myelination in the central regions, related to the somatosensory cortex, present already at birth. Myelination of the visual pathways occurs rapidly postnatally, and slow myelination, continuing until at least the age of 12 months is seen in the prefrontal white matter and in the parietal white matter. Newer, more sensitive methods show that myelination progresses at a slow pace until late childhood or adolescence. Klingberg and colleagues (1999), using diffusion tensor imaging (Chapter 3), found that the central white matter of the frontal lobes continues to myelinate after the age of 10 years. The corpus callosum is another structure with late myelination, which appears to continue into the young adult years (Pujol et al., 1993). Callosal volume, as measured on MRI scans, increases throughout childhood and adolescence, until at least the age of 18 years. Curiously, the rostrum and genu (anterior parts) of the corpus callosum appear to reach adult size earlier than the posterior corpus callosum (splenium)(Giedd et al., 1996).

Myelination in subcortical white matter and synaptogenesis in the overlying cerebral cortex appear to occur at about the same time, and both seem to follow the same hierarchical pattern. While some cortical function may be present in an unmyelinated system, it is probably limited, since unmyelinated central axons are unable to conduct the rapidly repeated depolarizations or bursts of activity that make up a message (Huttenlocher, 1970). A good correlation between myelination of the brain and function has been noted in children with delayed myelination, who also are likely to show a corresponding neurodevelopmental lag (Van der Knaap et al., 1991; Harbord et al., 1990). It appears that delayed myelination reflects the delayed development of neurons in the cerebral cortex, including the delayed growth of efferent (axonal) connections and of synapses.

### Relationships between Anatomy and Function

A major factor in the function of a given cortical area clearly is the origin of afferent connections and the destination of efferent connections. Several other properties also appear to be important for the function of a given cortical region. The size of a cortical region needs to be considered. A brain which has a large cortical region devoted to a specific function is likely to accomplish more complex analysis related to this function than a brain that has a smaller representation. Comparison of human cerebral cortex with that of other mammalian species shows the same basic cortical anatomy, with repetition of structurally similar vertical columns. However, in human cerebral cortex there is a marked increase in the number of these units, especially in cortical areas subserving complex functions, such as integration of information from different sensory inputs (association areas), language functions, reasoning, and the so-called executive functions. On the other hand, some cortical regions, such as, for example, the olfactory cortex, are decreased in size in the human brain when compared to those regions in several other species, reflecting the decreased importance of the sense of smell in the human. Cortical regions that are dedicated primarily to olfactory functions in other mammalian species have different functions in the human. For example, the hippocampus, a part of the olfactory system in lower mammals, has nonolfactory functions, including verbal and spatial memory, in humans (Abrahams, Pickering, Polkey, and Morris, 1997).

Another factor that appears to be important for cortical function is the size of the dendritic tree of individual cortical neurons. The dendritic complexity of pyramidal neurons is greatest in human cerebral cortex. For example, pyramidal cells in layer 3 of mature human prefrontal cortex have an average of 80,000 synapses per neuron versus about 10,000 in feline cerebral cortex (Cragg, 1975b). Within human cerebral cortex, the size of the dendritic trees of pyramidal cells relates to complexity of function. The total dendritic length of pyramidal cells at maturity is considerably less in the visual cortex than in the prefrontal cortex. Neurons in the visual cortex in general are smaller in size than neurons in the prefrontal cortex. The average number of synapses per neuron in layer 3 of adult human primary visual cortex is about 12,000, versus 80,000 in the prefrontal cortex (Huttenlocher, 1994). An interesting question arises: Is there a relationship between the adult size of the dendritic trees of neurons and the time that is available for the growth of dendrites? In the visual cortex, dendritic growth is completed some time between the ages of 6 and 24 months, while in the prefrontal cortex it is only about 50 percent complete by the age of 24 months. Perhaps the longer developmental period makes possible the development of the large dendritic trees required for the processing of complex information that occurs in human prefrontal cortex. Compared to the frontal cortex of other mammals, the human frontal cortex has the largest proportion of neuropil and the lowest neuronal density, reflecting the high order of dendritic, axonal, and synaptic growth. The human prefrontal cortex also has the slowest time course of cortical development, which extends over a period of at least 16 years. In general, the time required for brain development becomes shorter and the neuronal packing in the cerebral cortex becomes greater (neuropil, consisting of dendrites, synapses, axons, and glial fibers, makes up a smaller proportion of the total cortical volume) as one descends the phylogenetic scale (Ramon y Cajal, 1960).

Cerebral functions, including the intellect, may relate to the high complexity of dendritic growth. This is suggested by the correlation between diminished cortical dendritic growth and mental retardation in the human, as has been reported in a large number of mental retardation syndromes. These include phenylketonuria (Bauman and Kemper, 1982); cretinism, or congenital hypothyroidism (Eayrs, 1971); Rett syndrome, a genetically determined X-linked dominant disorder, seen mainly in girls, in which severe mental retardation is a prominent feature (Armstrong et al., 1995); and West syndrome, or infantile myoclonic seizures with mental retardation (Huttenlocher, 1974, 1991).

Conversely, increased environmental input and enriched learning ("environmental enrichment") have been associated with increase in the size and complexity of dendritic arbors (Volkmar and Greenough, 1972; Floeter and Greenough, 1979; Pysh and Weiss, 1979; Spinelli et al., 1980; Uylings and Parnavelas, 1981; Walsh, 1981; Green, Greenough, and Schlumpf, 1983; Greenough, Black, and Wallace, 1987).

Finally, the types of connections that are made between cortical neurons are important for normal function. A correct balance between excitatory and inhibitory synaptic inputs appears to be necessary. As yet we know rather little about the developmental aspects of excitatory (such as glutamatergic) and inhibitory (GABAnergic) synapses in human cerebral cortex. What is known is that immature neurons have the capacity to develop in several different directions, as far as neurotransmitters are concerned. The transmitter that is chosen by a developing neuron depends on the environment into which its axons grow. This has been especially well studied in the autonomic nervous system. Immature autonomic (sympathetic) neurons that normally synthesize norepinephrine as their neurotransmitter may instead develop into cholinergic neurons if redirected to innervate the heart, or if grown in tissue culture in a "conditioned medium" that contains an extract of cardiac muscle (Patterson, 1978).

While the formation of specific circuits that include axons, synapses, and dendrites is essential for the function of the cerebral cortex, a direct relationship between synapse number or density and cortical function is much less likely. If there were such a relationship, then a 3-year-old child's cortex should function better than an adult's (the child should be "smarter" than an adult). While it may be true that the 3-year-old has more potential for learning and intelligence than a given adult, this potential is as yet not realized, and may never be realized. The function of the cortex depends not so much on the number of connections as on the type of connections or functioning circuits that are formed. This is borne out by the study of synaptic density in brains from the mentally retarded, which often is found to be normal (Cragg, 1975a). At this point, we do not know why such brains do not function normally. However, findings from studies in developmental neurobiology provide important leads. Possible abnormalities include the mismatching of basic, genetically determined cortical connections (in particular failure of proper antegrade or retrograde signaling during synapto-genesis), defective dendritic growth, and defects in the choice of neuro-transmitters.

The discussion of structure-function relationships would be incomplete without a consideration of the relationship between growth of the neu-ropil, including synaptic development, and the onset of function in the im-mature cerebral cortex. The visual cortex lends itself especially well to such comparisons, since much is known both about its developmental anatomy and about the onset of its function. At birth, there is no clear indication of visual cortex function in human infants. While newborn infants can be shown to have some visual fixation and following, this may well be medi-ated by subcortical pathways. In contrast, functions that involve binocular interactions require processing at a cortical level, where extensive interac-tion of inputs from the two eyes first occurs. The emergence of visual func-tions that require bilateral interaction therefore is a measure of the onset of function in the primary visual cortex. These functions develop early in the first year of life, at the time of rapid synaptogenesis (Wilson, 1988). A good example is the development of stereopsis, the ability to see objects in three dimensions. This emerges very rapidly, at the exact time of rapid synapto-genesis in the primary visual cortex, between the ages of 3 and 5 months (Held, Birch, and Gwiazda, 1980; Teller, 1983; Wilson, 1988). It therefore appears that basic cortical functions develop early, at the time when synap-tic density in the cerebral cortex rises rapidly. This is followed by a much longer period of "fine tuning" of function that is partly environmentally controlled (Chapter 4) and that corresponds to the period when synaptic density is high, above adult levels (Wilson 1988). During the period of high synaptic density the visual cortex shows evidence of high plasticity, mani-fested, for example, by the reversal of strabismic amblyopia by patching of the good eye (VonNoorden and Crawford, 1979).

The findings in studies of the language areas lend support to the hypoth-esis that functional activity in a cortical area emerges at the time of rapid synaptogenesis. Basic language functions, comprehension and production of words and simple grammar, develop in the second year of life, earlier for receptive than for expressive language. Synaptogenesis occurs during the same period, somewhat earlier in Wernicke's area (speech perception) than in Broca's area (language production).

## Summary

Anatomical studies of human cerebral cortex development show that developmental changes occur over a remarkably long period of time, from the early embryonic period to at least mid-adolescence. The remodeling and pruning of dendritic and axonal branches and of synapses are important components of the development of the cerebral cortex. Many of the developmental events in the cerebral cortex are genetically determined. However, available evidence suggests that the early formation of synaptic contacts is to a large extent random. This leads to large numbers of synaptic contacts that are at first nonfunctional. Input to the cortex from sense organs and later from other cortical areas appears to be important for functional specification of early synapses. It is at this point that brain development becomes dependent on environmental influences. Synapses that are not used for functional circuits eventually disappear. The period during which there are large numbers of unspecified synapses may be one of increased ability of the cerebral cortex to process new information, and, as will be shown later, may represent an optimum period for certain types of learning. In the human, this period differs somewhat for different cortical regions. The human data, but not those from studies of subhuman primates, support a hierarchical pattern, with earlier development of primary sensory areas, followed by development of the motor and association cortex, with the latest development occurring in the prefrontal cortex, the cortical region that mediates the so-called higher cortical functions, including motivation, judgment, and reasoning.

# 3

## METHODS FOR THE STUDY OF FUNCTIONAL PLASTICITY

> In a group of normal children the cerebral blood flow and cerebral
> metabolic rate were found to be significantly higher than in normal
> young adults, and the cerebral vascular resistance was significantly
> lower.
>
> Charles Kennedy and Louis Sokoloff, 1957

Anatomical data, discussed in Chapter 1, provide information about the extent and limits of malleability in the development of the fine structure of the brain. They also suggest time windows, or critical periods, during which plasticity is likely to be high. They do not provide direct information concerning effects on cortical functions. For example, the demonstration that a change in environment affects dendritic growth does not tell us whether and to what extent the changes in the dendritic trees of cortical neurons relate to changes in brain function. Moreover, the anatomical methods provide only one data point for each subject. Longitudinal data, and especially data on the effects of interventions on developing cerebral cortex, are difficult to obtain by anatomical methods.

An exciting recent advance has been the development of methods by which the *function* of the brain can be assessed noninvasively and repeatedly. The effects of environmental factors on brain function can now be assessed in the same subject at multiple levels (metabolic, electrophysiologic, and cognitive). In this chapter, the available methods will be briefly described. They include: in vivo imaging of regional metabolism of the cerebral cortex by positron emission tomography (PET); magnetic resonance imaging of changes in cerebral blood flow related to the functional activity of the brain (fMRI); electrophysiologic (electroencephalogram, or EEG, and evoked potential) methods; magnetoencephalography and magnetic stimulation; computer simulations of developing neural networks; and cognitive and neuropsychological studies designed to separate environmental from inborn (genetic) influences on the development of cortical functions. The emphasis will be on the application of these methods to the study of plasticity. Readers interested in a more complete discussion of the physics and of the range of clinical applications related to developmental disorders of the functional imaging techniques should consult available reviews (for example, Shevell, 1999).

### Metabolic Studies of the Cerebral Cortex by PET

Convincing evidence that the metabolism of the brain in the child exceeds that of normal young adults was obtained more than 40 years ago in the pioneering work of Kennedy and Sokoloff (1957), which used a modification of the nitrous oxide method for measurement of cerebral blood flow (see the quote at the beginning of the chapter). Approximately half of the basal total oxygen consumption of the child is accounted for by cerebral metabolism. The original method provided the metabolic rate of the whole brain only. PET represents a considerable recent advance, since it yields a map of the regional metabolic activity of the brain as determined by the rate of glucose metabolism or by the rate of oxygen consumption. Under normal conditions, neurons are entirely dependent on glucose as the substrate for energy metabolism. The rate of glucose metabolism is a fairly accurate measure of energy consumption by cortical neurons. The metabolic activity of neurons, in turn, is thought to be closely linked to synaptic activity. Abolition of synaptic activity, for example by administration of a very high dose of barbiturates, is associated with a marked decrease in glucose consumption in the brain.

In the PET system, glucose metabolism is assessed by the intravenous injection of the radioactive tracer [18]F-fluorodeoxyglucose, which is taken up at the same rate as is glucose by metabolically active tissue. The regional concentration of the radioactive compound is measured by sensors that detect the emitted radioactive particles (positrons). More recently, oxygen consumption by brain tissue has been measured more directly by the use of oxygen[15] as the radioactive marker. Oxygen consumption is a somewhat

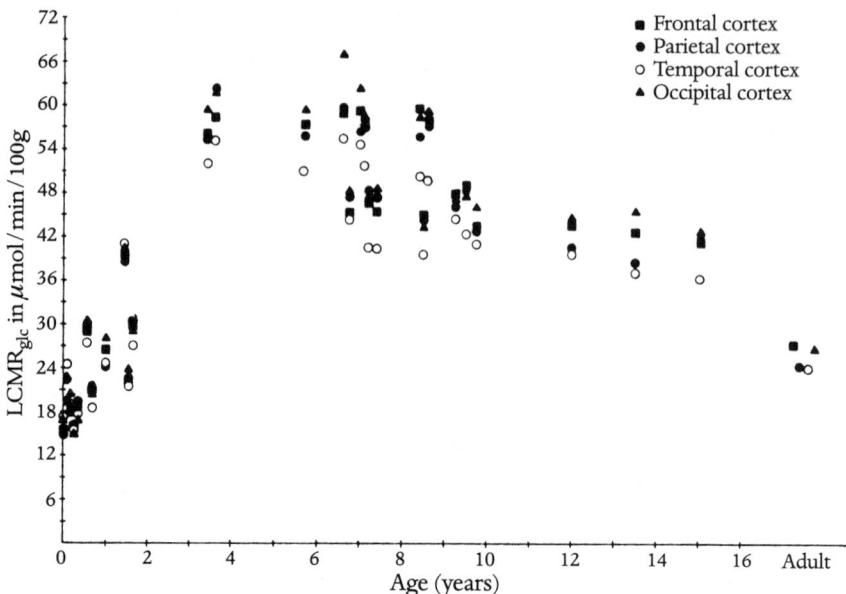

Figure 3.1. Regional cerebral metabolism for glucose as a function of age. (From Chugani, Phelps, and Mazziotta, 1987, fig. A, p. 492; reprinted with the permission of John Wiley & Sons, Inc.)

more exact measure of neuronal, and hence synaptic, activity than is glucose utilization. This is so because glucose metabolism includes a small component of anaerobic metabolism by glial cells. Glia are able to utilize the glycolytic pathway that metabolizes glucose to lactate, in the absence of oxygen. This anaerobic metabolism in glia is not measured in oxygen[15] scans.

PET can provide quantitative data concerning both the resting metabolic rate in specific cortical regions and changes in regional metabolic activity related to cortical activation by sensory stimulation, voluntary movement, and mental activity. The maturational changes in the resting cerebral metabolic rate have been studied by Chugani and Phelps (1986, 1990) and Chugani, Phelps, and Mazziotta (1987). They show remarkable similarity to the changes in cortical synaptic density that occur with maturation (Figure 3.1). There is a rapid increase in the cerebral metabolic rate for glucose in the first year of life, corresponding to the rapid increase in synaptic density. Cortical regions differ somewhat in the time course of metabolic activity increase. The prefrontal cortex lags about 5 months behind the more posterior cortical regions, with the phylogenetically newer dorsal prefrontal cortex the last to show metabolic increase, at about the age of 8 months (Chugani and Phelps, 1990). The metabolic data therefore confirm a hierarchical developmental program, with cortical regions mediating more complex cognitive functions the last to become metabolically active.

The resting metabolism of cerebral gray matter, as measured by PET, shows a plateau of a high metabolic rate, nearly twice the adult level, between late infancy and early adolescence, when cerebral metabolic activity declines to adult values. The PET data show a major decrease in cerebral energy metabolism in early adolescence, which probably reflects the synaptic pruning that occurs at that age. The combined synaptic density and PET findings suggest that an important change in the functional organization of the cerebral cortex occurs during adolescence.

The phase of rapid increase in metabolic activity corresponds to the onset of simple functions of a cortical area. Stereopsis, a function mediated by the primary visual cortex, develops at about the age of 3–4 months. Reaching for objects, a function mediated by the hand area of the motor cortex, emerges at 4–5 months, and performance of a simple memory task (the A not B task; Diamond 1985; Diamond and Goldman-Rakic, 1989), mediated by the dorsolateral prefrontal cortex, develops at 8–11 months. At the same time, it needs to be pointed out that the increase in metabolic activity, the rapid phase of synaptogenesis, and presumably the onset of function occur during infancy (prior to the age of 1 year) in all neocortical regions, even in the prefrontal cortex. However, the most complex functions, especially distinctly human ones such as abstract (including mathematical) thought, judgment, planning, and reasoning, appear to develop much more slowly and considerably later, during childhood and adolescence. There is no direct relationship between the development of these complex cortical functions on the one hand and the increase in cerebral metabolic rate or net synapse formation on the other. A number of these complex cortical functions appear to emerge during and after adolescence, that is, during the time of decline in metabolic activity of the cerebral cortex, when synaptic density decreases markedly, owing to synaptic pruning.

The functional data from PET do not show regional differences in the decline of the metabolic rate that occurs in early adolescence. In that respect they appear to differ from the synaptogenesis data. However, the synaptogenesis data so far obtained in human cerebral cortex show an early decline in synaptic density mainly in the primary sensory (visual and auditory) cortex, both of which are small regions lying to a large extent infolded below the cortical surface. They may therefore be difficult to image accurately with PET.

The use of PET in childhood has been limited by the invasiveness of the technique. PET requires intravenous infusion of a radioactive tracer, which is excreted by the kidneys and therefore is concentrated in the bladder. Repeated blood samples are needed. The exposure to radiation from a single scan has been decreased as the method has been improved, and it now is justifiable to use PET repeatedly and in studying a greater range of clinical problems than previously. However, study of normal children by this technique is rarely justifiable. Compared to fMRI, the method also has the drawback of poor temporal and spatial resolution. It is very expensive and requires a cyclotron to produce the very short half-life isotopes that are used.

A start has recently been made in the use of PET for assessment of the effects of early environmental changes on the development of brain function. Preliminary data have been presented on children with severe environmental deprivation during infancy, sustained by early placement into substandard orphanages in Romania. Children from these orphanages, adopted by U.S. parents, had decreased metabolism in several areas of the prefrontal cortex, including the cingulate gyrus, suggesting that the prefrontal cortex may be an area of the brain whose development is especially dependent on environmental stimulation (H. T. Chugani et al., 2001). Defective development of these cortical regions may form the basis of cognitive and behavioral abnormalities that have been ascribed to early environmental deprivation, such as the so-called cataclytic depression described by Spitz (1945) in infants raised in orphanages.

As yet, we have no data on the effects of enriched environments on cerebral metabolic activity. Such studies still are difficult to justify in humans. However, PET has been used to study plasticity in the metabolic activity of the cerebral cortex in response to nervous system damage or maldevelopment. A good example is the reaction of the visual cortex to the absence of afferent input from the eyes in children with congenital blindness. These children have high metabolic activity in the visual cortex, as high or higher than that in a seeing individual who is looking at an object. This is quite different from the effect of visual loss after maturation is complete, which leads to hypometabolism in the visual cortex. The functional significance of this difference is not immediately apparent from the available PET studies. However, recent studies using magnetic stimulation and evoked potential techniques suggest that, in children with early loss of vision, the visual cortex is recruited for other functions (see below and Chapter 4).

Another recent application of PET to the study of plasticity is related to the reaction of the immature brain to focal damage. PET—as well as other imaging techniques—has been used to study the reorganization of functional activity of the brain in children with early-acquired lesions (Muller et al., 1999). The results of these studies are presented in Chapters 5 and 6.

### Developmental Changes in Specific Neurotransmitter Systems: PET and Receptor Binding Studies

An exciting recent development in PET is the imaging of markers for specific neurotransmitters such as serotonin and dopamine. This puts into reach the study of effects of environmental stimuli on defined central pathways (such as the central serotonergic, cholinergic, or dopaminergic system). It also makes it possible to study the involvement of specific neurotransmitter systems in defined neurodevelopmental disorders. For example, D. C. Chugani, Muzik, Behen and their colleagues (1999) determined serotonin synthesis capacity in the cerebral cortex by using PET and the radiochemical alpha[$^{11}$C]methyl-L-tryptophan in subjects with autism and in sibling controls. Autistic children had a lower serotonin synthesis capacity than age-matched controls. In the controls, serotonin synthesis capacity reached a value more than twice the adult value by the age of 5 years and then decreased to the adult value. In contrast, serotonin synthesis capacity was low at the age of 5 years in autistic children and then gradually increased, reaching a value one and a half times normal by the age of 15 years (D. C. Chugani, Muzik, Behen, et al., 1999). These findings indicate the importance of developmental studies in individuals with neurodevelopmental disorders. In the case of the autism study, data obtained only in adolescents and adults would have provided potentially misleading results. Serotonin synthesis capacity depends at least in part on the number of serotonergic synapses. The findings therefore suggest that there is overproduction of serotonergic synapses early in childhood in normal subjects, that is to say that these synapses partake in the exuberant synaptogenesis that is characteristic of cortical synapses in general. The initial formation of serotonergic synapses appears to be decreased in the child with autism. The data furthermore suggest that later in development there is a decrease in the synaptic pruning of serotonergic synapses in the brain of the child with autism. The same sequence of developmental abnormalities—delayed early synaptogenesis, followed by a decrease in synaptic pruning—has been described in the primary visual cortex of animals deprived of normal formed visual images during the sensitive period (Chapter 2). The question therefore arises whether in the child with autism there may be decreased synaptic input to the neurons of the central serotonergic system. This system consists, to a large extent, of neurons in midline brain stem nuclei, the raphé nuclei, which project widely to forebrain structures, including the cerebral cortex.

Abnormalities of the central dopaminergic system have recently been described in subgroups of patients with "attention deficit disorder." Dougherty and colleagues (1999) reported overexpression of the dopamine transporter in a PET study of a group of adults with attention deficit/hyperactivity disorder.

The PET study of specific neurotransmitter systems holds considerable promise for important discoveries in other neurodevelopmental disorders, such as epilepsy (H. T. Chugani, daSilva, D. C. Chugani, et al., 1997), and specific language disorders, including dyslexia. These methods also will provide a new means for the assessment of treatments, including environmental manipulations, for these disorders. In other words, we may learn whether and to what extent environmental input can improve abnormalities in central neurotransmitters in children with specific cognitive disorders, such as infantile autism. This information may in turn be useful for the design of treatment protocols.

Prior to the development of PET, studies on neurotransmitter development in human cerebral cortex had been limited to receptor binding (ligand) assays and assays of neurotransmitter catabolic enzymes in postmortem cortical tissue. Kornhuber and colleagues (1988) studied binding of the ligand MK 801 to N-methyl D-aspartate (NMDA), one of the receptors at glutamatergic synapses. In human prefrontal cortex, they found an age-related curve of receptor binding, with rapid increase in the first year of life, a high plateau during childhood, during which receptor binding is about twice the adult value, and decline in late childhood, with adult values reached in adolescence. In a second study, Kornhuber and colleagues

(1988) found a similar age-related biphasic curve for glutamate binding. Similar developmental changes were observed by Diebler and colleagues (1979), who studied the development of glutamate decarboxylase, a marker for GABAnergic (usually inhibitory) synapses, in the frontal and parietal cortex. Johnston and colleagues (1985) found binding for a muscarinic cholinergic receptor in the prefrontal cortex to be higher in the 3-month-old than in the adult, suggesting that the cholinergic system may develop somewhat earlier than the glutamatergic and GABAnergic systems. Unfortunately, no later childhood ages were examined in the Johnston study. In general, the available data on neurotransmitter and receptor levels suggest that developmental overproduction followed by pruning occurs at all types of cortical synapses.

### Functional Magnetic Resonance Imaging (fMRI)

FMRI represents a major recent advance in the noninvasive study of cerebral function. It assesses changes in the cerebral metabolism indirectly, through their effect on the deoxyhemoglobin concentration in the blood that supplies a metabolically active cortical area. The increase in cerebral synaptic activity that underlies cortical information processing is accompanied by an increase in oxidative metabolism, which in turn leads to an increase in cerebral blood flow. The blood flow response characteristically shows an overshoot, beyond what is required for the maintenance of normal tissue oxygenation. As a result, the oxyhemoglobin concentration in regional blood vessels increases and the deoxyhemoglobin concentration decreases (Figure 3.2). The decrease in deoxyhemoglobin concentration results in a signal increase on MRI images (T2 weighted), corresponding to the region of increased cerebral metabolism and blood flow. Clusters of small volumes of tissue (voxels) with an increased signal are identified and are mapped onto brain slices or onto the cortical surface anatomy as determined by conventional MRI. For example, an fMRI of finger movements in the normal adult shows that activation occurs in the left precentral gyrus during right finger movement, and in the right precentral gyrus during movement of the left fingers; movement of the fingers of the left hand is also associated with a small ipsilateral activation. By contrast, an fMRI of finger movements after a left-sided stroke during the newborn period shows that movement of the paretic side is associated with bilateral activation in

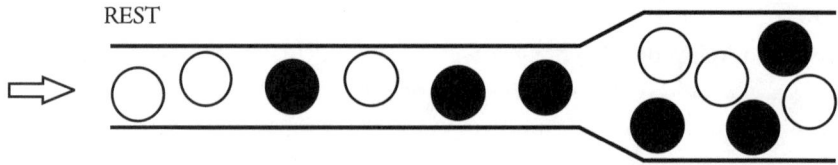

REST

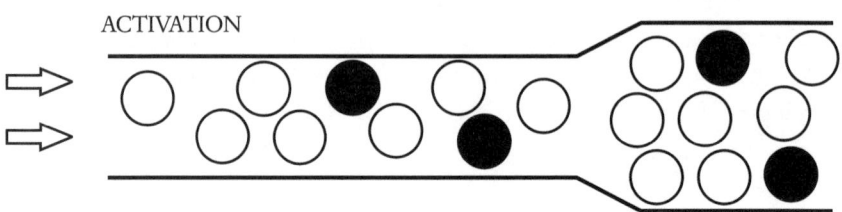

ACTIVATION

Figure 3.2. Schematic depiction of the basis of fMRI scanning. The method depends on the fact that blood flow increases in functionally active regions of the cerebral cortex. The increase in cerebral blood flow exceeds the oxygen consumption of the activated tissue. This results in dilution of the concentration of deoxygenated hemoglobin containing blood cells (filled circles) by the oxygenated hemoglobin-containing cells (open circles). (From Logan, 1999, fig. 1, p. 79; reprinted with the permission of the W. B. Saunders Company.)

the motor, premotor, and posterior parietal cortex; activation is more diffuse than in the normal individual (Cao et al., 1994). FMRI does not provide quantitative data on cerebral blood flow or metabolism. The technique requires the comparison of a scan during which the subject carries out a task (for example looks at the picture of a recognizable object) with a scan performed during a control period (for example, when the subject is looking at random scribbles), and calculation of the difference between signals (based on deoxyhemoglobin concentrations) during the task versus during the control conditions. In many tasks, a motor response, such as pressing a button, is required to confirm that a mental task has been completed (such as recognition of an object).

The fMRI method has a number of limitations. The selection of a proper resting or control condition for subtraction is very important and becomes more important and difficult as the complexity of the task that is assessed increases. No clear guidelines are available as to what constitutes an optimal baseline or control condition. Another drawback of the method is the fact that the region of increased cerebral blood flow may include feeding and draining vessels (arteries and veins) at some distance from the area of changed metabolic activity, which leads to some inaccuracy of functional localization. Finally, the noise of the magnets and the required confinement of the subject in a narrow tube make testing unacceptable for some subjects and difficult to perform on children. Nevertheless, the effects of complex mental tasks, including language tasks, have been assessed in children as young as 7 years old (Benson, Logan, et al., 1996; Hertz-Pannier et al., 1997; Logan, 1999). Practice in a simulator has been useful, as has positive reinforcement. The major advantage over PET is that the method is completely noninvasive and safe. No needle sticks are required and there is no exposure to ionizing radiation. Children can be studied repeatedly and their developmental trajectory can be followed. Another advantage of fMRI over PET is better spatial and temporal resolution. Finally, the technique is less expensive than PET. Widely available MRI equipment can be used, with minor modifications. However, at this time PET remains the "gold standard" of functional imaging, since it measures cerebral blood flow and metabolism much more directly than does fMRI. Validation of key findings on fMRI by PET study therefore is desirable. Where such comparisons have been made, similar results have been obtained with the two techniques (Ramsey et al., 1996).

The study of the effects of input to a given cortical area on functional localization is a field of intense, ongoing investigation. It has already yielded interesting new data on plasticity of the motor (Chapter 5) and language (Chapter 6) areas. PET has been used to assess the adjustment of the brain to early-acquired (usually perinatal) focal lesions. Presumably, such effects are long lasting, and can be studied after the child has matured, which avoids the limitations of fMRI in young children. The results need to be compared to the results with both normal control subjects and those with later-acquired lesions of comparable size and location. Remarkable reorganization of functional localization has been found after early-acquired focal brain lesions, including transfer of function to the normal hemisphere after unilateral brain damage. New data relevant to cerebral plasticity, based on fMRI, are accumulating at a remarkably high rate. Examples are discussed in Chapters 4, 5, and 6.

## Diffusion Tensor Imaging (DTI)

DTI is a newly developed magnetic resonance imaging technique that utilizes water as a source of image contrast, rather than hydrogen or proton density, as in conventional MRI, or deoxyhemoglobin, as in fMRI. DTI tracks the diffusion of water in brain tissue. This occurs randomly in all directions (isotropically) in the gray matter. In white matter tracts, the diffusion of water is anisotropic, parallel to the direction of the tract, since the axons that make up the tract inhibit lateral diffusion (see Beaulieu et al., 1999 for a recent review). The technique makes it possible to image white matter tracts in great detail. Unmyelinated as well as myelinated tracts are demonstrated, the latter more intensely. DTI has been used in developmental studies that show the growth of white matter systems. In the subcortical white matter, growth is greatest between birth and the age of 6 months, with further, slow growth between 6 months and adulthood (Nomura et al., 1994; Takeda et al., 1997). A study comparing DTI in prematurely born versus full-term infants showed delayed growth of subcortical fiber systems, probably due to delayed myelination, in the prematures (Huppi et al., 1998).

DTI is just beginning to be used for the study of pediatric populations. It holds considerable promise as a method for the investigation of reorganization of cortico-cortical and corticofugal connections in subjects with early brain damage.

## Electrophysiological Studies of the Maturation of the Cerebral Cortex

### ELECTROENCEPHALOGRAPHIC STUDIES

The electroencephalogram (EEG) is a recording of the electrical activity of cerebral cortex from multiple scalp electrodes. It shows characteristic waveforms, which vary with regard to both age and level of arousal. Its application to the study of cortical development and plasticity has been limited. An example of such a study was published by Feinberg and colleagues in 1990. These investigators analyzed the available ontogenetic data for the amplitude of delta (slow-wave) activity that is characteristic of sleep. This is low at birth, increases to nearly twice the adult value during the first 3 postnatal years, then maintains a high plateau for several years, and decreases to the

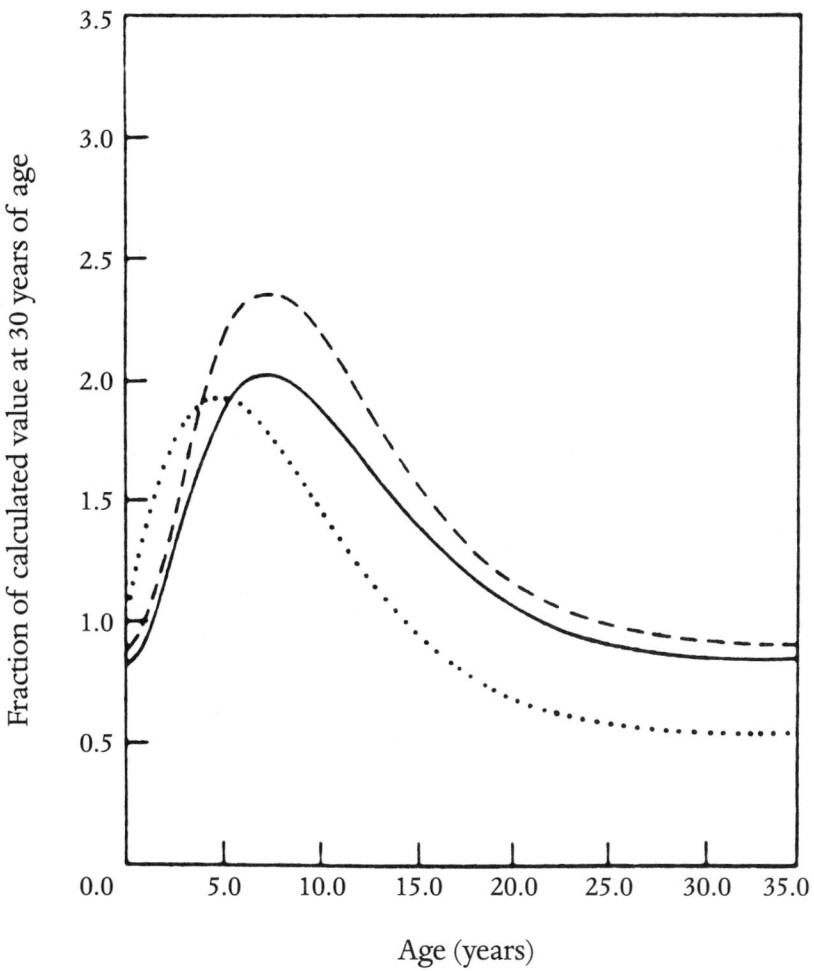

Figure 3.3. Synaptic density squared (dotted line), delta wave amplitude (dashed line), and cortical metabolic rate (solid line) as a function of age. The data were standardized to the value at 30 years. (From Feinberg, Thode, Chugani, and March, 1990, fig. 2, p. 153; reprinted with the permission of Dr. Irwin Feinberg.)

adult amplitude during adolescence. By now, this curve should be familiar, since it is very similar to the curve for age-related change in synaptic density, in the development of receptors for specific neurotransmitters and for regional cerebral metabolic rates (Figure 3.3). The increase in power of the resting EEG that is seen during childhood probably reflects increased neuronal activity related to the presence of exuberant synaptic connections.

The pruning of synapses that occurs during late childhood and adolescence is likely to have major effects on the organization of electrical activity in the cerebral cortex. A lower density of synapses, and the associated decrease in total neuronal activity, may increase the efficiency of signaling in the cerebral cortex. The amplitude of a signal, such as a sensory evoked response, is determined by the amplitude of the stimulus-evoked activity minus the amplitude of the background noise. Signaling may therefore be impaired by an increase in background noise related to the presence of excess synaptic connections.

Synaptic pruning may also account for the increase in the threshold for seizure activity that is seen during late childhood and adolescence. For example, simple febrile convulsions, that is to say seizures that are triggered by fever, are rarely seen after the age of 5 or 6 years, and several epilepsy syndromes such as absence (petit mal) seizures and benign childhood (Rolandic) seizures tend to disappear during early adolescence.

Synaptic pruning therefore may have both positive and negative effects. The positive ones may include a reduction in background noise and a decrease in the risk of uncontrolled neuronal activity, such as exists during convulsive seizures. The negative ones are likely to be related to the loss of unspecified synapses that would otherwise have been available for the construction of new cortical circuits.

### Evoked Potential Studies

Evoked potentials recorded from the scalp provide a measure of regional changes in cerebral electrical activity in response to specific sensory stimuli. The method is noninvasive, requiring only the pasting of recording electrodes with flexible leads to the scalp. It can be carried out at any age, including infancy. Even fetal recording has been attempted. The method as presently used has, however, rather poor spatial resolution, and functional localization is less accurate than that obtained by PET or fMRI. Stimuli have to be brief and novel to evoke a response, and usually an averaging technique is needed to yield consistent data.

A particularly useful response has been the event-related potential (ERP), which consists of a surface positive wave, about 300 milliseconds after the stimulus (a p300 wave) in mature cerebral cortex. The ERP is thought to be closely linked to cognitive processes. It occurs only when a reasoned response to a stimulus is required. The latency of the ERP is high in early childhood, and declines to adult values by about the age of 15 years (Courchesne, 1977; Goodin et al., 1978). This implies that information processing is slow in childhood, until early or mid-adolescence. This finding represents an additional piece of evidence favoring a major reorganization of cerebral cortex function in late childhood and early adolescence.

The use of evoked potentials for the study of cortical plasticity has been pioneered by Helen Neville. In a now classic study, Neville and colleagues looked at the cortical distribution of visual evoked potentials, comparing congenitally deaf with normal hearing subjects (Neville, Smith, and Kutas, 1983). The study represents the first clear demonstration of cross-modal plasticity in the cerebral cortex, namely the ability of a cortical area normally specified for one type of sensory input (auditory) to take on some of the functions of another sensory system (visual). The result has two interesting implications: First, cortical localization of functions is not entirely determined by genetic programs; a sensory cortex that has lost its normal input early in life is able to reorganize its function to some extent. Second, there is competition for cortical representation between sensory inputs. Competition for cortical representation and the resulting crowding out of some potential connections are themes in cortical development that repeatedly surface. They have been studied especially well in developing primary visual cortex (Chapter 4).

### Magnetoencephalography (MEG) and Transcranial Magnetic Stimulation (TMS)

MEG and TMS represent another new approach to the localization of cortical function. MEG makes use of the detection of magnetic fields that are set up by cortical electrical activity. Focal increase in neuronal activity leads to a change in the associated magnetic field, which is detected as a signal. The results may be confirmed by magnetic stimulation of the putative region of functional localization. For example, the hand area of the motor cortex might be localized by MEG as an area of change in a magnetic field related to hand movement or by another technique for functional localization, such as fMRI. The accuracy of the localization can then be checked by magnetic stimulation of the identified area on the scalp. The response to repetitive stimulation of the hand area consists of the contraction of muscles in the contralateral hand and fingers (Pascual-Leone et al., 1994). In most other cortical regions, repetitive transcranial magnetic stimulation (rTMS) results in negative effects. For example, rTMS of the motor speech area (Broca's area) results in the transient arrest of speech (Pascual-Leone, Gates, and Dhuna, 1991).

RTMS appears to be safe provided that the trains of stimuli are kept at or below 3 seconds, with stimulus intensity near the threshold for muscular contraction when the stimulus is applied to the Rolandic (motor cortex) region. More intense or prolonged stimulation carries a risk of seizure activity (Pascual-Leone et al., 1993; Chen et al., 1997; Wasserman, 1998). There are few animal studies addressing the issue of whether this form of cortical stimulation may cause tissue damage. The few histopathological data available for the human showed no evidence of cellular damage in a recently stimulated cortical area (Gates et al., 1992). The accuracy of cortical localization obtained with rTMS is superior to that obtained with the evoked potential techniques presently available. The technique is less demanding for the subject than is fMRI. For example, delineation of the motor cortex requires only that the subject lie quietly during the period of stimulation and recording. Such studies can therefore be carried out in children below the age for reliable performance of fMRI.

TMS holds considerable promise as a technique for the study of cortical plasticity (Pasqual-Leone et al., 1999). It can provide proof that a functional representation has been changed in response to focal injury to the brain or to abnormal sensory input. It has recently been used to study cross-modal plasticity in the visual cortex in subjects with congenital blindness (see Chapter 4).

### Comparison of the Imaging Methods

Several methods for the in vivo study of functional activity in the cerebral cortex are now available. Each has advantages and disadvantages that need to be considered when one chooses a method for a given study, and in the interpretation of results. Important properties of the most widely used techniques are summarized in Table 3.1. More than one technique may be appropriate for any given study. For example, for the study of cross-modal plasticity in sensory cortex deprived of its normal input, several techniques

*Table 3.1.* Comparison of the available imaging methods.

| Method | Temporal resolution | Spatial resolution | Need for patient cooperation | Potential risk |
|---|---|---|---|---|
| PET | Low (minutes) | Poor | Low | Radiation exposure |
| FMRI | Fair (seconds) | Good | High | None |
| MEG | High | High | Some | None |
| Mag. Stim. | Fair | High | Some | Convulsions |
| EEG/ERP | High (millisecs.) | Poor | Some | None |
| DTI | — | Good | Some | None |

can and have been applied. These have included PET, fMRI, magnetic stimulation, and evoked potential studies. In this case, results with one method such as PET have been validated and extended with those obtained by entirely different techniques, such as magnetic stimulation and evoked potential studies (Chapter 4).

### Assessment of Cognitive Function

Careful assessment of the various aspects of cognitive function including language, spatial, and mathematical ability is essential for the study of the plasticity of the developing human brain. Discrepancies in cognitive functions between children brought up in deprived and privileged environments have long been recognized. It has, however, been very difficult to separate the influence of genetic factors from environmental ones in such comparisons. Children brought up in enriched environments tend to have successful parents. Genetic endowment rather than upbringing may be the crucial factor in both the parents' and the children's high performance. A variety of methods have been tried to sort this out. Studies of identical twins reared in different environments have shown only modest effects ascribed to environmental differences. However, the differences in environmental stimulation in twin pairs tended to be small also, even when they were raised in separate homes, since out-of-home placements are often with relatives or with families that have qualities similar to those of the natural parents.

Recently, large long-term follow-up studies have been carried out to assess the effects of intensive enrichment programs in deprived populations. These studies require random assignment of children to control and enrichment groups and careful follow-up to adult life, to determine whether the effects of the program carry over into adult performance (Ramey and Campbell, 1984; Campbell and Ramey, 1994, 1995). Such studies are very expensive, and are becoming more difficult to carry out as community services for underprivileged children improve. The Abecedarian study of Ramey and colleagues showed large benefits for the enrichment groups. More recent replications have shown smaller and more variable results, especially in communities such as Boston, Massachusetts, that routinely provide high-quality intervention services for all underprivileged children. The Abecedarian study will be discussed in more detail in Chapter 6.

New techniques for the study of environmental effects on cognitive functions are needed, especially techniques that are relatively inexpensive and that can separate environmental from genetic factors. One such method is the longitudinal assessment of school effects on cognitive functions and on language development (J. Huttenlocher, Levine, and Vevea, 1998). In this method, children's cognitive growth is determined during a period when school is in session and the results are compared with progress made during the months of summer vacation. The method allows for separation of environmental (school) effects from effects due to other factors, including maturation and genetic programs (see Chapter 5). The problems related to recruitment of separate, properly matched control subjects are avoided, since the design involves comparison of performance at multiple time points in the same subjects.

### Computer Simulation of Developing Neural Networks

Computer simulations of the development of cortical connectivity provide interesting information concerning possible strategies for functional development used by biological systems (Edelman, 1987; Elman and Zipser, 1988; Elman, 1993; Elman, Bates, et al., 1996). Computer networks that begin with random connections are of special interest. They show that a functioning system can arise from such a network. A typical simulation of a developing neural network begins with a group of units each one of which is equally connected with every other one. Half of the units are excitatory, the other half inhibitory, both randomly distributed in the system. The computer simulation reproduces what is thought to occur during development of the cerebral cortex: initially, the system consists of randomly formed connections. The computer simulation then is provided with an input that consists of a variety of stimuli, which do have common and consistent elements. For example, the input may consist of shapes and lines, simulating visual input, or it may consist of speech sounds, simulating auditory input.

Jeffrey Elman used such a system to model language development. The computer network, exposed to speech sounds (words and word combinations), develops circuits of increased connection strength, representing recurrent sound patterns, usually single words. The first stage therefore corresponds to vocabulary learning. The computer subsequently learns to recognize word combinations and relationships between words, specifically simple grammar. A third phase is the gradual elimination of connections representing sound combinations that fail to occur in the input. The system therefore reproduces the likely steps in the development of functional activity in the language cortex. There is functional validation or specification of some of the initial random connections, stabilization of circuits that are used repeatedly, and weakening of connections that are not used, corresponding to synapse elimination. The computer learns better with some types of input than with others. The input should be simple, but varied. Repetition of the same stimulus is less effective than is an array of stimuli. Learning by the computer therefore has similarities to human learning. Computer simulations of functional development provide interesting clues as to the kind of cellular mechanism that may underlie learning in the brain of the infant.

### *Electrophysiologic Studies of Functional Development in Animal Models*

Physiological studies in humans, using the techniques described above, provide data concerning changes in the location and size of areas of functional representation, while cognitive assessments give information concerning the effects of environmental differences on mental functions. These techniques do not address the issue of the mechanisms involved in changes in cerebral functions related to environmental stimulation. The cellular events underlying cortical plasticity are best studied in animal models. The visual system has been studied most intensively. The classic work of Hubel, Wiesel, and their collaborators on the effects of visual input on the development of synaptic connections in the visual cortex represents the first and still most complete analysis of the cellular basis of cortical plasticity. This work will be summarized in Chapter 4.

Animal experiments have been important in defining the extent, time windows, and limits of cortical plasticity. In the study of lesion-induced plasticity it has become evident that effects of focal lesions on reorganization of the normal hemisphere need to be studied in adults as well as at various time points in development. The cerebral cortex appears to retain considerable capacity for reorganization in the adult (Chapter 8). It has, however, been difficult to study the effects of comparable lesions incurred at various ages in humans. Problems have included the difference in the reaction of brain tissue to focal damage in infancy versus adulthood and difficulty in matching infancy and adult cases with respect to the time interval between the occurrence of the lesion and the performance of the study. These problems have been addressed in animal studies, especially in the work of Jaime Villablanca and colleagues in the cat. This work shows persistence of plasticity in adult cerebral cortex, but to a lesser degree than in the infant's (Villablanca, Burgess, and Benedetti, 1986; Villablanca, Burgess, and Olmstead, 1986; Villablanca, Gomez-Pinilla, et al., 1987, 1988; Loopuijt et al., 1998).

### Summary

The study of functional plasticity of the cerebral cortex has been advanced greatly by the development of in vivo imaging techniques. For the study of normal variations and of the effects of sensory deprivation, functional MRI, magnetoencephalography, magnetic stimulation, and evoked potential methods have proven to be especially useful. Positron emission tomography has provided data largely with respect to lesion-induced plasticity. This method provides the most direct information concerning regional cerebral metabolism of glucose, brain oxygen consumption, and neurotransmitter status. It is, however, too invasive for use in the study of normal children. FMRI is emerging as the major method for studying the localization of cortical functions and their reorganization in response to focal lesions or to altered input. This is so especially in children and adolescents, where performance of PET studies is limited by the radiation exposure inherent to the method and by the inability of potential subjects to give informed consent.

Understanding of functional plasticity has also been aided by new experimental designs in cognitive psychology that make it possible to separate genetic from environmental effects on development. Finally, the classic electrophysiologic (single-unit) studies on the malleability of cerebral connectivity in response to changes in input, especially the work of Hubel and Wiesel on the development of neuronal activity in the cerebral cortex of the kitten, continue to provide a framework for the understanding of cortical plasticity at a cellular level.

# 4

## PLASTICITY IN SENSORY SYSTEMS

"Does the skin tell the somatosensory cortex how to construct a
map of the periphery?" We propose that the answer to the title's
question is yes.

Hendrik Van der Loos and Josef Dörfl, 1978

### Plasticity in the Primary Visual Cortex: Animal Studies

The anatomy and physiology of neural plasticity have been studied more
extensively in the visual cortex than in any other cortical region. This is so
because much is known about the normal anatomy, physiology, and function of the visual cortex, which makes it possible to study environmental
effects on development at various levels, and to correlate findings from multiple disciplines. Both the extent of environmental influences and the occurrence of genetic or innate developmental programs have been assessed in
the visual cortex. Strong influences of both have been found. The environmental effects are especially interesting, since the visual cortex subserves
very basic functions such as stereopsis and the processing of the shape and
movement of objects across the visual field. These are functions that develop in all organisms with normal eyes, visual pathways, and cortical anatomy, apparently independently of learning. One therefore might have expected that genetic influences would be predominant in the visual cortex.
Yet it has been found that environmental influences are important in the development and maintenance of synaptic connections, even in this cortical
region. The classic work was carried out in animal studies. A considerable
body of evidence points to similar effects in humans.

Work on the development of the visual cortex in kittens and that on the effects of visual deprivation was pioneered by Hubel and Wiesel (1963a, b). It is beautifully summarized in the Nobel lecture delivered by Torsten Wiesel in Stockholm in December of 1981 (Wiesel, 1982). The initial approach was single unit recording from neurons in the visual cortex of kittens. At birth, some cortical neurons already respond to visual stimuli such as movement of a slanted bar across the visual field. These responses are weak and inconsistent, and they lack the remarkable specificity of activation by visual stimuli seen in mature visual cortex. The results show that, at birth, there already are functional synaptic connections in the lateral geniculate nucleus and between axonal branches of the geniculate neurons and neurons in layer 4 of the visual cortex. Some units in other cortical layers could also be activated, which suggests that some intracortical circuits are already in place. These connections must have formed in the absence of visual stimuli, but not necessarily in the absence of input to the visual cortex from the retina, which is known to generate waves of activity spontaneously prior to birth (Stryker and Harris, 1986).

During the first 3 or 4 postnatal weeks, there is a progressive increase in units that can be activated by visual stimuli, and by the age of 4 weeks adult patterns of neuronal activity are reached. The responses become stimulus specific, that is, a given neuron becomes activated when the test stimulus is presented in one orientation only. Most neurons respond to stimuli from one eye only. In layer 4c, the cortical layer that receives direct synaptic input from the lateral geniculate nucleus, groups of neurons that are activated by stimulation of the retina of one eye alternate with groups of cells activated by the other eye. These alternating patches are referred to as ocular dominance columns. Following their discovery by Hubel and Wiesel (1963a) by single cell recording, these ocular dominance columns were also demonstrated anatomically by several methods, including unilateral intraocular injection of $^3$H-proline. This molecule is carried from the retina to the visual cortex by antegrade, transneuronal transport. Its location in the visual cortex can then be demonstrated by autoradiography, which shows patches of radioactivity alternating with patches of tissue without the radioactive label in layer 4c of area 17, marking alternating inputs to the visual cortex from the injected and the opposite eye (Hubel, Wiesel, and Stryker, 1978; LeVay, Stryker, and Shatz, 1978).

## THE EFFECTS OF BILATERAL VISUAL DEPRIVATION

After birth, visual input becomes important for the further development and maintenance of cortical connections. In the kitten, bilateral deprivation of formed visual images by eyelid suture or by placement of opaque lenses from about the age of 3 weeks to the age of 8 weeks results in relatively subtle changes in the response of cortical neurons to visual stimuli. Responses are less stimulus specific, and are seen more frequently to stimuli from both eyes. The ocular dominance columns are less clearly defined. Perhaps most important, many neurons fail to respond to any visual stimuli, although they are "spontaneously" active (Wiesel and Hubel, 1965, Singer and Tretter, 1976). This raises an intriguing question: What kinds of connections are made by these visual cortex neurons that cannot be activated by visual stimuli? Might they have been recruited for functions other than visual? Behaviorally, these animals appear blind when the opaque lenses are removed. This finding differs markedly from the results obtained in adult animals, where deprivation of formed visual images by placement of opaque lenses or by eyelid suture has no effect on visual competence. However, some degree of recovery of vision over time does occur even when bilateral visual deprivation extends from birth to maturity.

Interesting effects have also been reported in response to abnormal visual input (partial deprivation). For example, kittens raised with goggles that expose one eye to vertical stripes and the other to horizontal stripes develop an abnormal distribution of orientation-selective neurons in the visual cortex. Kittens that have worn such goggles from birth to the age of 12 weeks have a change in orientation-selective cortical neurons that corresponds to the direction of the stripes on the goggles. Cortical neurons responding to a vertical bar projected on the retina of the eye exposed to the vertical stripes are increased, as are neurons responding to a horizontal bar projected on the retina of the eye exposed to horizontal stripes (Leventhal and Hirsch, 1975; Stryker and Sherk, 1975). In the normal kitten, vertical and horizontal bars produce more cortical activation than slanted bars. This difference is probably in part due to the predominance of vertical and horizontal lines over slanted ones in normal visual experience, and in part to an intrinsic developmental program. Early exposure to slanted stripes will increase the number of cortical neurons that respond to diagonal bars at the same angle as the stripes to which the eye was exposed, but many neurons with response to horizontal and vertical orientation also persist (Leventhal and Hirsch, 1975). These experiments suggest that what the kitten sees—and what very likely we see—is to a significant extent determined by visual experience early in life. In addition, there are groups of neurons that develop connections independent of visual input, perhaps in response to a genetic program. The important point is that visual experience early in life has effects on the synaptic organization and function of the visual cortex. As we will find out later, the same is true for the auditory system. Our ability to discriminate sounds depends on exposure to specific sound patterns during a critical period in infancy.

### The Effects of Unilateral Visual Deprivation

The most striking neurophysiologic and anatomical changes occur in relation to asymmetric visual input (Hubel and Wiesel, 1970). In the kitten, unilateral eyelid suture or unilateral placement of an opaque lens that allows diffuse light stimulation, but precludes formed visual images, results in an increase in neurons in the visual cortex that respond to visual stimuli presented to the seeing eye. At the same time, there is a marked decrease in neurons that respond to stimuli presented to the occluded eye. This effect occurs during a "sensitive period" or "critical period" that begins at the postnatal age of 3 weeks and declines after the age of 8 weeks in the kitten. Unilateral visual deprivation during the critical period also leads to striking anatomical changes, both in kittens (Olson and Freeman, 1980) and in nonhuman primates (Hubel, Wiesel, and LeVay, 1977; LeVay, Wiesel, and Hubel, 1980). There is shrinkage of the ocular dominance columns with input from the occluded eye and expansion of the columns related to the seeing eye. The experiments in kittens fitted with opaque lenses or subjected to unilateral eyelid suture clearly show environmental effects on the formation or maintenance (or both) of the ocular dominance columns: unilateral visual deprivation, applied during the sensitive period, leads to abnormalities of the synaptic organization of the visual cortex. Synaptic connections from the seeing eye to the visual cortex become strengthened, while connections from the occluded eye decrease. No such effect is seen when occlusive lenses are applied to mature animals.

Unilateral visual deprivation in infant monkeys has effects similar to those in kittens. The effects on neuronal activity in the primary visual cortex are largely reversible if a period of visual deprivation of up to 90 days, starting at the age of 30 days, is followed by an equal period of monocular deprivation of the other eye (Crawford et al., 1989; Harwerth et al., 1989). However, these animals have an almost complete absence of bilaterally driven neurons, indicating failure of recovery of binocular interactions and of stereopsis. This result resembles findings in humans with monocular deprivation of formed vision in early infancy (see below).

## THE EFFECTS OF INDUCED DYSCONJUGATE GAZE

Effects similar to those obtained by occlusion of one eye have been seen following the induction of dysconjugate gaze by section of the lateral rectus muscle or by removal of its motor innervation (Hubel and Wiesel, 1965). Lateral rectus paralysis does not restrict visual input. It does remove the ability to fuse the images from the two eyes. The long-term effect depends on the age at which the lateral rectus muscle section is carried out: muscle section prior to the end of the sensitive period results in rapid suppression of the image from the squinting (adducted) eye. Loss of connections occurs between the squinting eye and the visual cortex, and there is unilateral shrinkage of the ocular dominance columns. No such effects occur if the muscle section is carried out after the sensitive period, that is to say after the age of 8 weeks in the kitten.

The anatomical and functional changes produced by unilateral lateral rectus muscle section are at least partly reversible. Recovery of vision takes place if the animal is forced to use the adducted eye by occlusion of the preferred (normally moving) eye, prior to the end of the sensitive period. The anatomical and functional changes become permanent if this therapy is delayed until after the end of the sensitive period.

## COMPARISON OF RESULTS IN KITTENS AND IN MONKEYS

Experiments in monkeys have shown both similarities between and differences from those in kittens (Hubel, Wiesel, and LeVay, 1977; LeVay, Wiesel, and Hubel, 1980). In the monkey, development of the cerebral cortex is more advanced at birth than it is in the kitten. Ocular dominance columns are already well formed in monkeys at birth, which indicates that this process can occur in the absence of visual stimulation, and therefore is at least in part independent of environmental control. It may, however, require some type of afferent input from the eyes. As already mentioned, Stryker and Harris (1986) have found retinal electrical activity in the form of waves of depolarization, which are present prior to any light exposure. This spontaneous retinal activity may be important for the early development of the cerebral cortex, including the ocular dominance columns. No clear conclusion about genetic versus environmental factors in the initial formation of ocular dominance columns can be drawn from the monkey data. While it is possible that there is a genetic program for their development, it appears just as likely that their development depends on afferent input to the visual cortex related to spontaneous retinal activity.

The visual cortex of the monkey—like that of the kitten—has a critical period, during which the synaptic organization is sensitive to visual deprivation, both unilateral and bilateral. However, this critical period starts early in the monkey, at birth or at the age of 1 week at the latest (Horton and Hocking, 1997). The deprivation effects begin to decline between the ages of 6 and 10 weeks, are still detectable at the age of 1 year, but are absent in the adult (Wiesel, 1981; Horton and Hocking, 1997). The data underscore an important fact, namely that critical periods tend to end gradually. In many cases, such as in the plasticity of human language areas, some degree of malleability in response to input persists in the adult.

### MECHANISMS OF PLASTICITY IN THE VISUAL CORTEX

The work of Wiesel and Hubel proves that plasticity in the formation and maintenance of connections in the visual system is related to cortical rather than subcortical mechanisms. There is no evidence of retinal damage related to the placement of opaque lenses, eyelid suture, or lateral rectus muscle section. The responses of the lateral geniculate nucleus—the way station in the pathway from the retina to the visual cortex—to visual stimuli are not affected (Wiesel, 1981). These findings are in agreement with the view that plasticity is a property of the cerebral cortex that is shared little if at all by subcortical structures. The only effect that is seen in lateral geniculate neurons is a decrease in the size of neurons that receive their innervation from the deprived eye. This effect probably relates to competition for synaptic sites in the visual cortex. In this competition, the input from the occluded or from the squinting eye loses out, and most of the connections are made with the dominant eye. The smaller field of cortical innervation is reflected by a decrease in the complexity and number of branchings of axons made by geniculate neurons when they reach layer 4 of the primary visual cortex. Decrease in size of the lateral geniculate neurons is not seen when deprivation of vision is bilateral. Smaller size therefore is not likely to be secondary to decreased neuronal activity related to the absence of formed visual images. The size change that occurs in these neurons underscores a point made in Chapter 1, namely that the size of cortical neurons relates to their connections—their input as well as their efferent pathways. These in turn can be modified by the environment, as is proven by data on the developing visual system.

There is some evidence that different functional systems in the visual cortex may have different susceptibilities to monocular deprivation. Two distinct functional systems have been described in the visual system: the magnocellular system, which projects to the parietal cortex and mediates spatial information (where is the object?), and the parvocellular system, which projects to the temporal lobe and has to do with object recognition (what is the object?). The parvocellular system appears to be more affected by deprivation of visual forms, and the critical period may extend longer than in the magnocellular system, which indicates that different functional circuits may be more or less environmentally regulated (LeVay et al., 1980; Horton and Hocking, 1997).

### Plasticity in the Primary Visual Cortex: Observations in Humans

Observations in human infants and children with visual disorders as well as studies of postmortem visual cortex have shown many parallels with the animal data. They show similar susceptibility of the immature human visual system to sensory input, and indicate the existence of critical periods.

#### THE EFFECTS OF BILATERAL VISUAL DEPRIVATION

Data concerning early visual deprivation have been collected from patients with congenital cataracts. These infants have naturally occurring opaque lenses that preclude perception of formed visual images. Removal of such cataracts, if postponed to the adult years, does not restore normal vision. There is some perception of shape and color, but this cannot be incorporated in such a way as to provide useful information. Instead, these stimuli seem to interfere with normal functions and are experienced as having negative effects to the point where patients have been unable to cope with the sudden influx of sensory stimuli that they were unable to interpret (Taylor et al., 1979). These negative effects are avoided if the congenital cataracts are removed in infancy. Visual acuity and interocular alignment tend to be normal if congenital cataracts are removed prior to the age of 8 weeks (Mohindra et al., 1983). Infants with a 4-month or greater delay in removal of the cataracts have visual acuity that is no better than that of a normal newborn. Such infants have rapid improvement in visual acuity after cataract removal, indicating that visual input is necessary for the postnatal development of vision in humans. Recovery tends to be incomplete, with a residual permanent decrease in visual acuity and with defective binocular interaction, manifested by strabismus (Taylor et al., 1979; Mohindra et al., 1983; Maurer and Lewis, 1993; Maurer, Lewis, Brent, and Levin, 1999). Vision is permanently and severely impaired if removal of congenital cataracts is not carried out until the adult years. The data suggest that during the first 8 weeks or so of life development of the visual cortex progresses normally or nearly so in the absence of exposure to formed visual images. Visual stimuli seem to become important for the development and maintenance of visual functions after the age of 8 weeks, namely at about the time of rapid synaptogenesis in the visual cortex in the human.

Observations made of premature infants support the conclusion that the presence of formed visual input during the peri- and early postnatal period has little or no effect on the development of the visual cortex. Premature infants have exposure to visual stimuli prior to the normal time of birth. Yet they show no apparent acceleration of visual development. The maturation of visual evoked potentials in prematures proceeds at the same pace as that of infants born at term (Harding et al., 1989; Pietro et al., 1997). The development of visual alertness (fixation and following of visual stimuli) is no more advanced in prematures at POD 40 weeks than it is in full-term neonates. Observations made of monkeys delivered prematurely show no effect on the time course of synaptogenesis, supporting the view that early developmental events, including the onset of rapid synaptogenesis, are not stimulus sensitive (Bourgeois and Rakic, 1989).

The brain of the monkey is more mature at birth than is that of the human newborn. Cortical anatomy and function in subhuman primates at birth differ from those of human neonates. The period of rapid synaptogenesis is already well under way in the monkey at birth, whereas in the human it does not begin until about the age of 2 months. The critical period for impairment of vision following visual deprivation starts at or near birth in the monkey, while in the human there appears to be a "grace period" of at least 8 weeks postnatally, during which visual deprivation is tolerated without permanent impairment.

## The Effects of Unilateral Visual Deprivation

In human infants, as in kittens and in monkeys, the effects of unilateral deprivation tend to be more severe than those of transient bilateral deprivation. The data on humans derive from the study of infants with unilateral congenital cataract (Maurer and Lewis, 1993). None of these children had normal visual acuity in the affected eye, and visual acuity above 20/160 was achieved only when the cataract was removed before the age of 5 months, and the operation was followed by prolonged occlusion (patching) of the normal eye. Removal of the unilateral cataract not followed by patching results in strabismus and in strabismic amblyopia in the affected eye.

### Strabismus

The data concerning the effects of the environment on the development of human visual functions derive to a large extent from the study of children with strabismus (squint), a condition in which normal binocular interactions fail to develop. The eye movements are dysconjugate, usually with one eye turned inward ("convergent strabismus," or esotropia). There is, however, no eye muscle paralysis, and each eye moves fully when tested with the opposite eye closed. Squint, or nonparalytic strabismus, as seen in human infants resembles the condition produced when lateral rectus muscle section is carried out in kittens, as was done by Hubel and Wiesel. In both conditions, the eyes are dysconjugate and there is inturning of the abnormal eye. However, strabismus in the human infant differs from that in the kitten, in that the eye that turns in can be shown to abduct fully when each eye is tested separately. In human infantile strabismus the defect lies in the binocular interaction, not in the function of the eye muscles. The infant with strabismus often develops a dominant eye for fixation. In that case, the squinting eye looses vision (amblyopia). Amblyopia can be prevented and even reversed if the child is forced to use the squinting eye for part of the day, by placement of a patch on the dominant eye or by the use of glasses that blur the image on the dominant side. Reversal of amblyopia is possible up to a critical age of 6–7 years (VonNoorden and Crawford, 1979; Assaf, 1982). After that age, amblyopia is permanent. The effect of strabismus or of other forms of dysconjugate eye movements—such as lateral rectus palsy—incurred after the age of 7 years differs from that of early-acquired or congenital squint. Patients with late-acquired dysconjugate eye movements develop diplopia (double vision), a rather disabling symptom, which is permanent.

The critical period for strabismic amblyopia in humans is much longer than is that of other mammals, as expected from the generally slower time course of human cortical development. This results in a larger time window during which amblyopia can be reversed by patching than is seen in other mammals.

The cellular basis of strabismic amblyopia in humans is unknown. However, it is likely to be similar to that in other mammals, namely competition for synaptic sites, leading to asymmetric input from the lateral geniculate nucleus to the visual cortex, and shrinkage of the ocular dominance columns related to the squinting eye. The existence of ocular dominance columns in human visual cortex has been demonstrated (Hitchcock and Hickey, 1980; Horton and Hedley-White, 1984), but there have been no studies of these in brains from individuals with strabismic amblyopia.

It is of interest that the upper age limit for reversal of strabismic amblyopia occurs at about the age when synapse elimination nears completion in human visual cortex. This finding lends some support to the view that the presence of exuberant (unspecified) synapses may be important for the synaptic plasticity that underlies the reversal of amblyopia. A schema of a possible relationship between exuberant synapses and the formation and plasticity of ocular dominance columns is provided in Figure 4.1. This schema assumes that the formation of random connections between the right and left geniculate axon terminals and the dendrites of layer 4 cortical neurons is an early event. This is followed by selective synapse elimination, which results in balanced input from the two eyes, represented by alternating inputs from one and the other eye, the ocular dominance columns. This balance is disturbed in strabismus, but can be restored if the child is made to use the squinting eye prior to the end of synapse elimination.

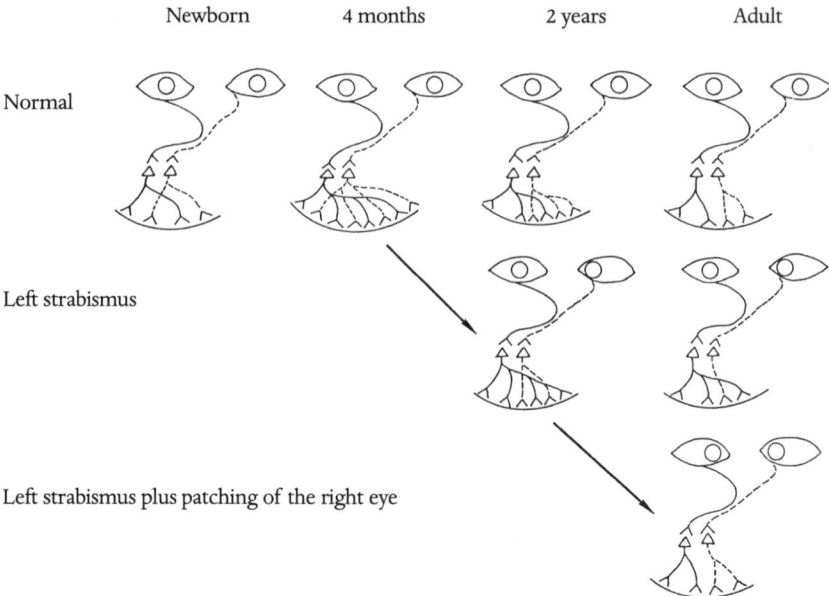

Figure 4.1. Schematic representation of normal synaptogenesis in layer 4c of the calcarine cortex, the effects of strabismus, and the effects of strabismus with patching of the dominant eye (see the text). (From Asbury, McKhann, and McDonald, 1992, fig. 6-2, p. 66; reprinted with the permission of the W. B. Saunders Company.)

## COMPETITION FOR SYNAPTIC SITES: NEGATIVE ASPECTS OF PLASTICITY

The data on the effects of monocular visual deprivation during infancy illustrate a possible negative aspect of neural plasticity, namely competition for synaptic sites, where there can be losers as well as winners. Strabismic amblyopia is of limited benefit to the child. It prevents diplopia, which can be a very bothersome and disabling symptom, but does so at the expense of unilateral blindness. The blindness appears to be due to competition for synaptic sites, where synapses that would normally have made geniculo-cortical connections with the squinting eye are replaced by connections from the dominant eye. The dominant eye invades the areas of representation of the squinting eye. Asymmetric visual attention, not the absence of retinal stimulation or the presence of dysconjugate gaze per se, leads to amblyopia. We know this from the study of children with strabismus who learn to use both eyes alternately (alternating strabismus). Amblyopia does not occur in these children, nor does diplopia. Unlike patients with adult-acquired strabismus, children with alternating congenital strabismus learn to suppress the image that is transmitted from whichever eye is squinting.

Competition for synaptic sites may also play a role in the adjustment of the brain to focal injury. In the adult, focal brain injury tends to lead to specific deficits related to the known function of the tissue that was damaged, such as aphasia in a patient with a left temporal lobe lesion, constructional apraxia with a right parietal lobe lesion, and so on. Focal brain damage in the infant does not lead to such specific deficits. Most functions are preserved to some extent, but each tends to be carried out somewhat less well. The result often is a general decrease in the efficiency of information processing. This effect may be due to competition for synaptic sites, leading to the invasion or taking over of areas of representation by functions whose normal areas of representation have been destroyed. This possible mechanism, referred to as the crowding hypothesis, will be further discussed in the chapter on plasticity in language functions.

### Reorganization of Function in Visual Cortex Deprived of Its Normal Input: Cross-modal Plasticity

The ability of immature sensory cortex to reorganize itself anatomically may, under certain circumstances, have positive effects on function. This has been shown to occur when visual input is absent from early life, specifically in the congenitally blind. The cerebral cortex of the blind loses almost all its afferent input, and might be predicted to have low synaptic and metabolic activity. This indeed is the case when blindness is acquired during adult life. However, the reaction of the visual cortex in congenital blindness is very different. Veraart and colleagues (1990), using PET, made an interesting discovery when they compared the glucose utilization of the visual cortex in subjects with congenital and adult-acquired blindness. The metabolic rate for glucose in the visual cortex of the congenitally blind was high, similar to that of the visual cortex in a normal, seeing subject during visual exploration. This was in marked contrast to the visual cortex in subjects with adult-acquired blindness, where metabolic activity was low, as expected, similar to that of normal visual cortex when the eyes are closed. This finding raised the question whether visual cortex deprived of afferent input early in life might become specified for some new function, not related to vision. At the single unit recording level, this would be expected to result in many cortical neurons that are "spontaneously" active (activated by stimuli that are not under the control of the experimenter), but cannot be activated by visual stimuli. Indeed, several investigators have reported such cells in animals deprived of formed visual images during the critical period (Wiesel and Hubel, 1965; Buisseret and Imbert, 1976; Singer and Tretter, 1976; Blakemore and von Sluyter, 1975).

A follow-up PET study by a group at the University of Louvain (De Volder et al., 1997) measured cerebral blood flow, oxygen consumption, and glucose utilization in the visual cortex of patients with early-acquired blindness and in blindfolded adult volunteers. All three parameters were increased in the visual cortex of the early-blind group, which suggests increased metabolic activity in the visual cortex. This is thought to be largely due to increased synaptic activity, related either to an increased discharge rate or to the increased density of normally active synapses. There is some evidence that both of these may occur: Bourgeois and Rakic (1996) found an increased proportion of asymmetric (considered to be excitatory) synapses in the visual cortex of early-enucleated monkeys. These extra excitatory synapses may represent the morphological basis for increased neuronal activity in deafferented visual cortex. A decrease in the normally occurring synapse elimination has been reported in kittens deprived of formed visual images by eyelid suture during the critical period (Cragg, 1975c; Winfield, 1981). This may provide an explanation for the finding that ocular dominance columns are less distinct in congenitally blind than in normal animals.

There is increasing evidence that these changes in early deafferented visual cortex may be of functional significance. Kujala, Alho, and their colleagues (1990) and Alho and colleagues (1993) reported a study utilizing event-related potentials in which auditory responses were found in the visual cortex of the early blind. Several recent studies have found evidence of somatosensory processing in the visual cortex of subjects with early blindness. The possible enlistment of the visual cortex for Braille reading is supported by PET studies and by magnetic stimulation. PET has shown activation of the visual cortex during Braille reading in blind subjects (Sadato et al., 1996). In a very interesting application of transcranial magnetic stimulation of the brain, Cohen, Celnik, Pascual-Leone, and their colleagues (1997) found arrest of Braille reading during magnetic stimulation of the visual cortex in congenitally blind subjects. In contrast, magnetic stimulation of the visual cortex in control subjects had no effect on tactile performance. Evidence for cross-modal plasticity has also been found in the congenitally deaf, where visual information processing appears to expand into cortical regions normally specified for auditory functions (Neville, Kutas, and Schmidt, 1982). This form of plasticity, utilizing sensory cortex that is deafferented, appears to be common (Rauschecker, 1995). It may underlie the special skills acquired by persons with impairment of sensory input in one modality, such as Braille reading and improved ability to navigate in space in the congenitally blind (Schlaggar and O'Leary, 1991).

Cross-modal plasticity appears to depend on the stabilization of normally transient inputs to the sensory cortex and/or on the formation of aberrant inputs after loss of the normal afferent connections (Frost, 1982, 1990). Several examples of such inputs have been described. These include transient connections between auditory and visual cortex in kittens (Innocenti and Clarke, 1984), which may become stabilized and functional after removal of input to the visual cortex from the eyes in the neonatal period. Transient connections between the somatosensory system and the visual cortex have also been described. Stabilization of these connections may underlie the ability of the visual cortex to process Braille reading in the congenitally blind human (Asanuma and Stanfield, 1990) (see Figure 4.2).

Binocular deprivation of vision from birth has effects on cortical areas outside the visual cortex, including sensory association areas in the parietal cortex. In visually deprived cats, inputs from the visual cortex to the suprasylvian association cortex are replaced by increased auditory and somatosensory inputs. Spatial tuning of auditory neurons is increased. Behavioral testing of these animals has shown improved sound localization, compared to that of control animals, suggesting that these adaptations, observed at a single unit level, are functionally significant (Rauschecker, 1995a,b).

### Plasticity in the Auditory Cortex

Knowledge concerning plasticity in immature auditory cortex is much more limited than is that for the visual cortex. We do know that, like the visual cortex, the auditory cortex receives bilateral sensory input. However, the bilateral interactions are much more complicated. In the visual system, afferent fibers from each eye to the visual cortex do not have bilateral con-

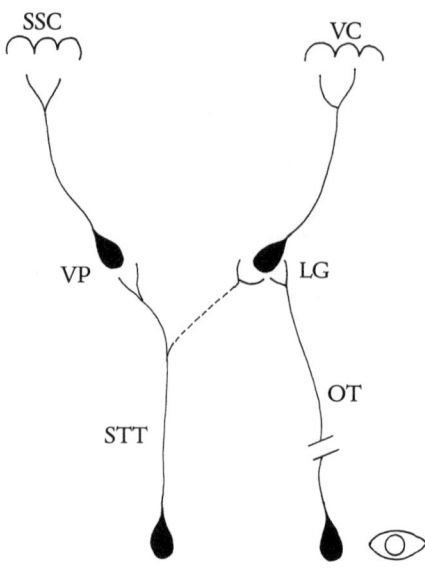

Figure 4.2. Aberrant innervation of the visual system by somatosensory axons (dashed line) in the congenitally blind mouse. STT = spinothalamic tract; VP = nucleus ventralis posterior of the thalamus; SSC = somatosensory cortex; OT = optic tract; LG = lateral geniculate nucleus; VC = visual cortex. (Drawing by Arun S. Dabholkar.)

nections until they reach the thalamus (the lateral geniculate nucleus), and even in the geniculo-calcarine fibers, input from the two eyes is largely separate. The first major interaction with input from the opposite eye occurs at the level of primary visual cortex. This arrangement may lead to competition between cortical inputs from the two sides, as has been discussed earlier in this chapter. In contrast, interactions between left and right inputs to the auditory cortex occur at multiple levels of the brain stem, including the medulla (via bilateral connections to the superior olivary nuclei) and the pons (via Probst's commissure, which connects the nuclei of the lateral lemniscus). There is much bilateral processing already in the brain stem, related to sound localization. A simple competition between left and right auditory inputs at the cortical level is therefore less likely.

### Long-Term Effects of Hearing Loss in Infancy

Unilateral hearing loss is quite common in infancy and early childhood, usually secondary to chronic or recurrent otitis. This does not lead to permanent unilateral deafness; that is to say, the effect is not analogous to that of unilateral deprivation of input in the visual system. Rather, there is return of normal hearing after the chronic infection and fluid accumulation in the middle ear has been corrected, unless damage to the middle or inner ear has occurred.

There has been concern that transient bilateral hearing impairment during a critical early childhood period might permanently impair auditory information processing. Children with a history of chronic bilateral otitis media during infancy, especially during the first 6–12 months, were found to have delayed language development at the age of 1 year (Wallace et al., 1988) and 3 years (Teele et al., 1984). However, retesting of such children at school age showed no difference between language function in children with a history of chronic otitis in infancy and controls, suggesting that there is significant ability to catch up (Schilder et al., 1993). Maw, Wilks, Harvey, and their colleagues (1999) found initial language delay in 9-month-old infants in whom treatment of bilateral otitis media with effusion by insertion of drainage tubes had been delayed, when compared to those with early intervention. However, this difference was no longer apparent at the age of 18 months. These follow-up studies have not identified a "time window," during which hearing impairment leads to permanent deficits in auditory information processing. However, subtle permanent deficits are not ruled out by the studies now available.

### THE EFFECTS OF COCHLEAR IMPLANTS

Recent data suggest that permanent age-related deficits in the ability to recover auditory functions do occur after prolonged, severe hearing loss that begins in infancy. These data derive from the results obtained with cochlear implants. In this procedure, electrodes are implanted into the cochlea, through which electrical signals from a microphone are transmitted to the inner ear. The electrical stimuli activate any remaining auditory nerve fibers. At least some auditory nerve cells (hair cells) have to be intact for the device to be of benefit. From the point of view of plasticity in the developing auditory system, the results obtained in congenitally deaf individuals have been quite interesting. Congenitally deaf children with cochlear implants learn oral language better than age-matched congenitally deaf children who had not had such therapy (Tyler, Fryauf-Bertschy, et al., 1997; Tomblin et al., 1999). Results are especially good in young children, under the age of 3 years (Tyler, Gantz, et al., 1997; Tomblin et al., 1998). The benefits from cochlear implants in the congenitally deaf gradually decrease with increasing age at the time of implantation until adolescence, at which age the implant is no longer effective (Fryauf-Bertschy et al., 1997; Bonn 1998). Several questions remain; data concerning the "time window" for the effectiveness of cochlear implants are limited. It is not known whether the inability to learn language after cochlear implants by adults and teenagers is due to developmental changes in primary auditory cortex or in the language areas (Broca's and Wernicke's areas). Adults with congenital deafness have had a longer period of language deprivation than children, and duration of the lack of exposure to speech sounds rather than a decrease in language learning capacity in the adult may be a factor. In the case of deafness acquired after establishment of language, the time interval between loss of hearing and insertion of the cochlear implant appears to be an important factor, with greater benefit when implants are performed soon after loss of hearing.

In the auditory cortex, as in the visual cortex, there is evidence that early loss of normal input may lead to processing of information from other sensory modalities. Neville and colleagues (1982, 1983) compared visual evoked potentials in normal adults and in adults with congenital deafness. The deaf subjects had larger visual evoked responses than the normals in all brain areas, including the temporal lobes near the auditory cortex. This effect was found to be age limited. It was not seen in persons that became deaf after the age of 4 years (Neville, Schmidt, and Kutas, 1983; Neville and Lawson, 1987). This form of cross-modal plasticity may be mediated by transient connections between the auditory and visual systems, as has been observed in cats and hamsters (Dehae, Bullier, and Kennedy, 1984; Frost, 1984; Innocenti and Clarke, 1984). The availability of more cortical circuits for visual information processing may underlie the ability of the congenitally deaf to learn sign language and lip reading.

### Plasticity in the Somatosensory System

Sensory representation in the cerebral cortex exists as a map of the body surface on the somatosensory cortex in the postcentral gyrus (Figure 4.3). This map shows a close resemblance to the well-known "homunculus" of representation of muscle groups in the motor cortex (the precentral gyrus). Touch stimulation of a small patch of skin leads to activation of neurons in the area of representation in the somatosensory cortex and to inhibition of neuronal activity in the surrounding cortical areas.

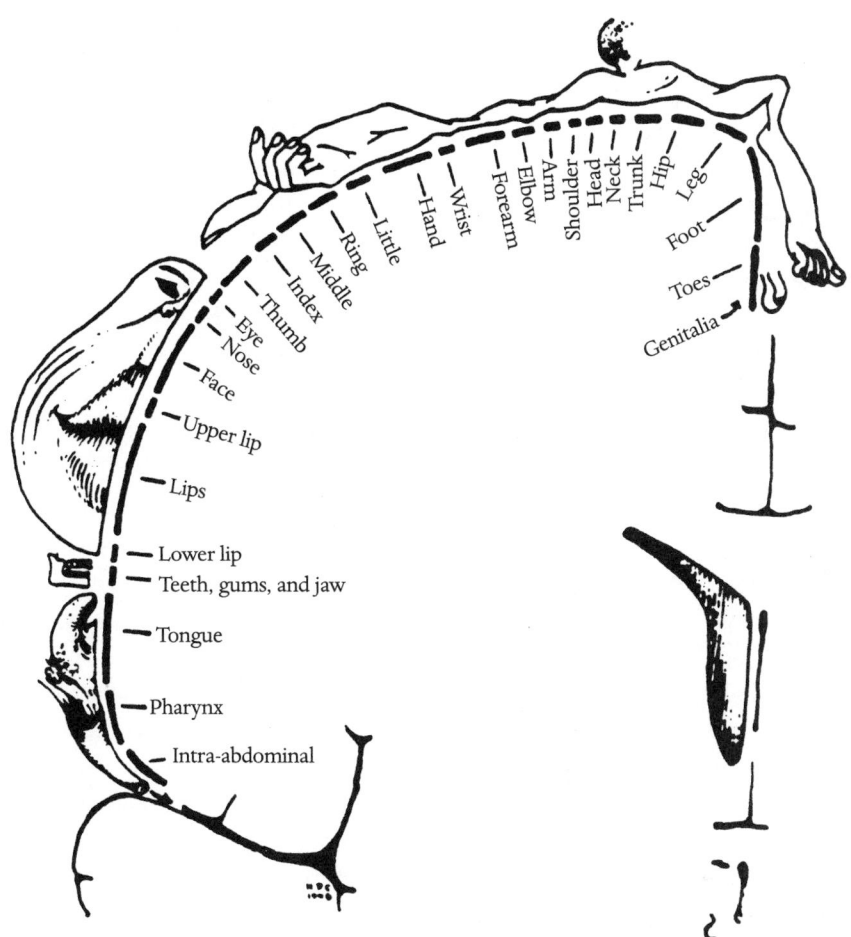

Figure 4.3. Homunculus of sensory representation in human postcentral (somatosensory) cortex. (From *The Cerebral Cortex of Man* by W. Penfield and T. Rasmussen, fig. 114, p. 214, MacMillan, 1950, copyright, 1950, by The Macmillan Company; reprinted with the permission of the Gale Group.)

## PLASTICITY IN CORTICAL SOMATOSENSORY MAPS IN MONKEYS

The map of the body in the somatosensory cortex is subject to changes in the periphery, indicative of considerable plasticity in the somatosensory systems. For example, without plasticity, amputation of a digit would be expected to lead to an absence of peripheral input via thalamocortical afferent fibers to the region of the somatosensory cortex that represents the removed digit. It has been shown that this is not the case (Merzenich et al., 1984; Buonomano and Merzenich ,1998). In fact, the cortical representation of neighboring, intact fingers expands into the area vacated by the amputated digit. This occurs within 2 months of amputation and has been interpreted as due to the sprouting of thalamocortical axons in neighboring cortical regions into the area that has lost its peripheral input.

More recently, Pons and colleagues (1991) reported a more massive reorganization of sensory representation in macaque monkeys after complete deafferentation of a forelimb. The cortical region that had received input from the deafferented forelimb now responded to stimulation of the skin of the face on the side of the deafferentation, a region whose cortical representation is contiguous with the normal forelimb representation on the somatosensory cortex homunculus.

Plasticity in somatosensory maps is an important example of recovery after injury that persists in adult life, and is therefore described in more detail in Chapter 8, on adult plasticity.

### PLASTICITY IN THE BARREL CORTEX OF RODENTS

A beautifully studied example of developmental plasticity is found in the barrel system of the mouse. Rodents have very prominent vibrissae, which they utilize as they navigate through space. The sensory input from the vibrissae is transmitted through the very large rodent trigeminal nerve to the thalamus, and from there to a specialized region of the somatosensory cortex, the barrel field. The name derives from the cellular anatomy of this region, which consists of collections of neurons in the shape of a barrel. The normal rodent has one barrel for each peripheral whisker. More remarkable, the early removal of whiskers or transplantation of extra whiskers causes the cortex to produce either fewer or more than the normal number of barrels, in order to maintain one barrel for each whisker. There is a "time window" for this form of plasticity, with an upper age of about 7 days after birth. This is the approximate age at which silent synapses disappear from rodent barrel cortex, which supports their possible role in some types of functional plasticity (Isaac et al., 1997). These experiments suggest that there is a fixed program in the developing barrel system that specifies one barrel per whisker, but that the actual number of barrels that is formed is environmentally determined, by the number of whiskers that are present during a critical period of cortical development. The barrel field arranges itself to accommodate the number of barrels that are required. Barrels tend to be smaller in rodents with a larger number of whiskers, larger when there are only a few whiskers (Van der Loos and Dörfl, 1978).

Van der Loos and associates found that genetic strains of mice with different numbers of whiskers have correspondingly different numbers of barrels. They then bred mice selectively to produce animals with larger numbers of whiskers than are seen in wild strains (Van der Loos, Welker, and Dörfl, 1986). These animals not only had the correct number of barrels (one barrel per whisker), but had enlargement of the entire barrel field in the sensory cortex (Welker and Van der Loos, 1986). In other words, "the skin tells the somatosensory cortex how to construct a map of the periphery" (Van der Loos and Dörfl, 1978).

These studies provide proof of the importance of sensory input to the cerebral cortex for the development of cortical anatomy and function. Changes in afferent stimulation during development can cause changes in the structure and function of the brain that persist in the adult. Finally, the results of the breeding experiments show that genetic variations in sense organs or sensory receptors may lead to modification of input to the brain, which in turn may affect the size and complexity of cortical areas (Killackey, 1990). Here we have an instance of plasticity in the structure of the cerebral cortex in relation to genetic change in the periphery, a startling interplay of genetics and environment in brain development.

An example of cross-modal plasticity involving vibrissal somatosensory cortex has been reported in cats. Kittens deprived of vision since birth develop hypertrophy of the facial vibrissae. In association with this, there is enlargement of the vibrissal cortex. Enlargement of this cortical area may underlie the increased use of the vibrissae for navigation in space in visually deprived animals (Rauschecker, 1995a).

### THE PLASTICITY OF SOMATOSENSORY MAPS IN HUMANS

#### Evidence from Phantom Limb Sensations

Evidence of reorganization in the somatosensory cortex in response to amputation of a limb has been reported in adult humans by Ramachandran and associates (1993). These investigators found that humans with upper limb amputation often had sensory experiences related to the missing limb when the face on the side of the amputation was stimulated. Similar sensations were reported on stimulation of the skin of the stump proximal to the line of amputation. Stimulation of these skin areas tended to reproduce phantom limb sensations. These sensory changes occurred less than 4 weeks after amputation (Ramachandran, 1993). Such rapid rearrangement of functional localization suggests a mechanism other than axonal sprouting. One possibility is the unmasking of connections between the deafferented cortex and the surrounding regions of the sensory cortex that were previously suppressed by inhibitory activity. Such unmasking of synaptic connections was reported in 1977 by Patrick Wall, who found enlargement of the receptive fields of neurons in the afferent spino-thalamic system immediately after cold block of a sensory nerve. Phantom limb sensations after amputation may be an example of the major role that the inhibitory (GABAnergic) system has in the functional specification of both subcortical and cortical neurons. For example, in the somatosensory cortex, a localized sensory stimulus leads to activation of a small pool of neurons, and to inhibition of neuronal activity in a much larger, surrounding area (the inhibitory fringe)(Kuffler, 1953; Mountcastle and Darian-Smith, 1968). Both the focal activation and the surrounding inhibition disappear after removal of the sensory input by deafferentation. Removal of the inhibition may lead to the strengthening of weak connections from regions that represent neighboring areas of skin. These connections were always present, but were obscured by inhibitory activity. Removal of the primary sensory input may not necessarily lead to remapping by the growth of new connections. Reorganization of the cortical sensory map may be related to changes in synaptic efficacy. The findings suggest that the cortical sensory map is not a fixed entity, but rather that it is dependent on the input that the cortex receives from the periphery. This type of plasticity, related to the activation of nonfunctional circuits, persists throughout the life span and is an important mechanism for recovery from focal lesions in the adult (Chapter 8). The cellular mechanisms for modulation of synaptic efficacy have recently been worked out, most notably by Kandel and colleagues (Kandel, Schwartz, and Jessell, 1991).

### Evidence from Patients with Early Lesions of the Somatosensory Cortex

Evidence for reorganization of the cortical somatosensory map in humans derives from the study of subjects with perinatal focal brain damage. FMRI study during sensory stimulation (stroking of the affected hand) has shown activation in the ipsilateral somatosensory cortex (the postcentral gyrus) in three out of nine subjects with perinatal damage to one hemisphere (Chu, Huttenlocher, Levin, and Towle, 2000). In the remaining subjects, there was either persistence of activation in the damaged hemisphere ($N = 3$) or no detectable cortical activation ($N = 3$). The latter was not seen in normal young adult controls. Absence of activation was not associated with increased severity of sensory dysfunction, which suggests that processing of sensory input may have occurred at a subcortical level. It therefore appears that the developing brain may have several possible strategies for functional reorganization after injury to the somatosensory cortex.

### Summary

Evidence of remarkable plasticity related to changes in sensory input has been found in sensory systems—starting with the work of Hubel and Wiesel in the visual system and more recently with studies of the auditory and especially the somatosensory systems. In the visual system, a classic example of stimulus-related plasticity relates to conditions in which there is asymmetric input from the two eyes. The cellular localization of this plasticity appears to lie in primary visual (calcarine) cortex, more specifically in cells of layer 4c that receive input from the lateral geniculate nuclei. In the auditory and somatosensory systems subcortical mechanisms may also be important, since these systems already have widespread connections at brain stem and thalamic levels, providing the anatomical basis for interactions between different inputs. Negative as well as positive effects related to changes in input are observed. What is clear is that the sensory systems are not fixed structures with immutable connections, even in the adult. While there is much that is genetically determined, these systems, laid down in rough outline by genetic programs, become malleable by the environment soon after birth, and can be modified significantly by the input they receive.

In the visual cortex, both negative and positive effects are seen in relation to visual input. An example of a negative effect is the loss of vision in the squinting eye or in the eye deprived of visual input, if the deprivation occurs during the sensitive period. An example of a benefit derived from neural plasticity in the visual cortex is the reversal of strabismic amblyopia, when the child is forced to use the squinting eye while still in the sensitive period. Both effects are related to competition between input from the two eyes to the visual cortex. In this competition the connections from the normal eye take over synaptic sites that normally would have been occupied by connections from the squinting eye. Competition between inputs is a mechanism of plasticity that is characteristic of the visual cortex and that is not seen to a similar extent in other cortical areas for which information is available.

We know rather little concerning plasticity in the auditory cortex. There is some evidence from evoked potential studies that areas of the cortex specified for auditory information processing may be enlisted for the processing of visual information in the congenitally deaf. There also is limited evidence for a "time window" during which restoration of hearing is likely to lead to the emergence of speech in children who became deaf prior to the normal age of language learning. This "time window" appears to extend from infancy to early adolescence. However, further data are needed to confirm this.

Considerable data concerning plasticity are available for the somatosensory cortex. There is evidence that the functional specification of large cortical regions can be changed by amputation or deafferentation of limbs. In the vibrissal system of rodents, there is evidence of actual structural changes in the sensory cortex in response to removal or transplantation of vibrissae. The effects of transplantation of vibrissae on the structure of sensory cortex have a "time window," ending only a few days after birth. This form of plasticity may be mediated by the utilization of functionally unspecified, silent synapses. Plasticity related to changes in the sensory maps in response to loss of afferent input from local skin regions persists in the adult. A number of different strategies appear to be used, including the sprouting of axon collaterals from neurons in neighboring cortical areas into the region that has lost its input and disinhibition of dormant connections. Developing—and to a certain extent adult—cerebral cortex, deprived of its normal inputs, does not become inactive, but instead seeks input from other systems. These changes in input may lead to improved function, or may give rise to negative symptoms such as phantom limb sensations.

# 5

## PLASTICITY IN THE MOTOR CORTEX

> In mammals, comparable lesions affecting motor status have far less
> permanent and severe effects when the injury is sustained in infancy
> than when it occurs in adulthood.
>
> Margaret Kennard, 1942

### The Effects of Perinatal Lesions in Animal Models

Next to the visual system, the motor cortex has been studied most thoroughly with regard to the capacity to reorganize. In contrast to the visual system, where most of the evidence derives from experiments and clinical observations related to differences in input, the motor cortex has yielded information on malleability almost entirely from studies of the effects of focal cortical lesions. We are here dealing with a somewhat different form of plasticity, the ability of the brain to reorganize after injury, referred to as "lesion-induced plasticity." The questions asked are: What are the effects of unilateral motor cortex ablation on motor function? Do these effects differ in relation to the age at which the lesion is made? Is recovery better when the lesion occurs early in life, that is, is there a "window of opportunity"? What changes, if any, are there in the functional representation of movements in the motor cortex? Is there reorganization of the remaining motor cortex that leads to strengthening of ipsilateral connections? Are new connections formed?

Lesion-induced plasticity in the motor cortex has been studied extensively in both animals and humans. In both, the studies have been concerned with the effects of focal lesions that include the motor cortex, and with the effects of hemispherectomy (removal of an entire cerebral hemisphere). Neonatal lesions have been studied most extensively.

## A Historical Note

The question whether recovery from a focal cortical lesion is superior if the lesion occurs early, during the perinatal period rather than in the adult, was first investigated in motor cortex in the 1930s by Margaret Kennard at Yale (Kennard, 1940, 1942). This work was carried out in monkeys in whom the motor cortex on one side was surgically excised in early infancy. The recovery of infant monkeys was compared to that in animals with comparable adult lesions.Unlike adult monkeys, infant monkeys recovered very rapidly and learned to walk and climb normally. This effect has been referred to as the "Kennard Principle" (Teuber, 1974). Kennard did notice that even in infant monkeys, recovery was incomplete. In particular, the monkeys with cortical ablations in infancy did not grasp objects as well with the contralateral hand as did normal monkeys.

## The Effects of Perinatal Lesions in Monkeys

Kennard's results were challenged by Passingham and colleagues (1983), who pointed out that Kennard's adult control monkeys were studied and sacrificed a few weeks after surgery, while the infants were allowed to survive for up to 2 years. The question therefore arose whether the differences in motor function observed between adult and infant lesions are due to the different lengths of the recovery period rather than to the superior ability of the immature brain to undergo reorganization of function. These investigators therefore repeated the Kennard experiments with appropriate adult controls, in which the time from cortical ablation to study was equal to that of monkeys with neonatal lesions. They confirmed Kennard's results that infant monkeys recover more rapidly. However, the adult monkeys that were operated on also showed remarkable, if slower, improvement, and were eventually able to run and climb without limping. Improvement in contralateral hand function was limited in both groups. Neither the monkeys operated on in infancy nor the monkeys with adult lesions were able to perform skilled fine finger movements, such as a thumb-forefinger grasp. It was concluded that early acquisition of a lesion does not improve eventual functional recovery over what is seen in adult animals. The fast recovery of the infant monkeys was explained on the basis of subcortical control of simple voluntary motor activities, such as walking and climbing, in the infant animals, although no direct evidence of this was cited. If walking and climbing in monkeys is mediated subcortically, one would expect them to be spared in bilateral ablations of the motor cortex. This, however, is not the case. Passingham and colleagues did point out an important aspect of lesions made during infancy: disability becomes apparent gradually. In particular, the inability of operated on infant monkeys to perform skilled hand and finger movements did not become noticeable until the age when these functions normally emerge.

Unfortunately, the design of the Passingham study is flawed as well, especially if one's interest lies in the effects of perinatal lesions. The infant monkeys had removal of motor cortex at various ages, with a range of 7 to 89 days, and no attempt was made to analyze the data by age at the time of surgery. The cerebral cortex of the monkey is already quite mature at birth, when compared to that of other species, including rodents, cats, and humans. Even the newborn monkey brain therefore is not a good model in which to study the likely effects of cortical lesions in human neonates. By the age of 2–3 months the monkey brain is highly developed, with the exception of synaptic pruning, which continues until about the age of 2 years. The results obtained in lesions made in a 2- or 3-month-old monkey therefore have limited relevance to the study of plasticity related to perinatal brain lesions in humans. As will be pointed out later in this chapter, enhanced recovery of motor functions after perinatal lesions has been convincingly demonstrated in species with late brain development, especially rodents and felines. What the studies in monkeys do show is that following ablation of the motor cortex, contralateral hand and finger movements are most severely impaired independently of the age at the time of the lesion. In that respect, the findings in monkeys are in agreement with observations in humans.

The limited recovery of function after perinatal lesions in the hand area of the cerebral cortex may relate to certain unique developmental features of the primate pyramidal motor system. Armand and colleagues (1997) studied the development of connections between cortico-spinal axons and spinal motoneurons, by antegrade axonal transport of horseradish peroxidase (HRP), injected into the motor cortex (hand area) in monkeys. They found that cortico-spinal axons had reached their appropriate level in the spinal cord at birth, but that they did not yet terminate on anterior horn cells in the spinal cord gray matter. Cortical projection on spinal motoneurons developed late, in the first 5 postnatal months. During the period from birth to the age of 3 years there was no evidence of exuberant cortico-spinal projection, followed by withdrawal, as occurs in many other projection systems. If a transient cortico-spinal projection occurs in the monkey, it must therefore be a prenatal event that is completed by the newborn period, and could not be a factor in recovery from a perinatal lesion.

The development of the pyramidal motor system in primates differs from that in rodents, in that rodents have transient cortico-spinal axons that persist in the postnatal period and that may be a factor in their recovery of motor functions after neonatal unilateral ablation of the motor cortex (see below). Another major difference is the fact that cortico-spinal axons in primates make direct synaptic contact with anterior horn cells (motoneurons) in the spinal cord, whereas they terminate in the intermediate gray matter in rodents. The direct connections between cortico-spinal axon terminals and spinal motoneurons make it possible to transmit a message from the motor cortex to muscles at remarkable speed. At the same time, this very exactly determined system may have less plasticity than systems that are less strictly specified for a given function. In this regard, it is of interest that in humans, plasticity in the motor system is greatest in facial movements, in a system that retains the more primitive pattern of lack of direct connections between cortico-bulbar axons and cranial nerve motoneurons (see below).

Anatomical studies of the monkey brains with postnatal cortical resections have failed to show any changes in the cortico-spinal pathway originating in the remaining intact cerebral hemisphere (Sloper et al., 1983). Specifically, there was no evidence of any increase in uncrossed (ipsilateral) cortico-spinal projections. The monkeys with infancy period cortical resections were found to have a small, crossed cortico-rubral pathway, originating in the remaining intact motor cortex, and terminating in the magnocellular portion of the red nucleus. Such a pathway is not found in normal monkeys, which indicates that some sprouting of new connections takes place even when a cortical lesion is made relatively late in development. A large crossed cortico-rubral tract has also been found in kittens after neonatal hemispherectomy, in contrast to normal kittens, which have few crossed cortico-rubral fibers (Fisher et al., 1988). The functional significance of this new tract is unknown. The finding is, however, of interest since the red nucleus receives input from the cerebellum and is the origin of a large efferent system to the spinal cord, the rubro-spinal tract. The cerebellum has important motor functions, which act to refine and smoothen voluntary motor activities. The cortico-rubral connections may therefore represent an attempt to substitute for the cortical motor deficit by increasing cerebellar influences on motor control.

### Post-Lesion Neural Plasticity in Rodents

Recovery from focal motor lesions has been studied most extensively in rodents. In the rat, the pyramidal motor system is normally entirely crossed. The axons of pyramidal neurons in layer 5 of the motor cortex travel ipsilaterally via the pyramidal tract to the medulla oblongata, where they cross over to the opposite side in the decussation of the pyramids. The pyramidal tract axons then descend in the contralateral spinal cord, where they make synaptic contacts with neurons located in the posterolateral spinal cord, rather than directly with anterior horn cells. This system is only partly developed in the rat at birth. Pyramidal tract axons are still growing into the spinal cord, and may not yet have reached their place of synaptic contact.

There is evidence indicating that initially more pyramidal neurons send axons to the spinal cord than persist in the adult. These transient fibers include a component that descends into the ipsilateral spinal cord. These ipsilateral fibers apparently do not make synaptic contact with spinal cord neurons under normal conditions (Joosten et al., 1987). They are withdrawn by about postnatal day 14, at which age the development of the rat brain is near the level of the human infant's brain at birth. Some of these transient axons originate in cortical areas outside normal motor cortex, even in occipital (visual) cortex (D'Amato and Hicks, 1978; Nah, Ong, and Leong, 1980; Stanfield, O'Leary, and Fricks, 1982; Leong, 1983; C. H. Bates and Killackey, 1984; Stanfield and O'Leary, 1985b). The cells of origin of the transient cortico-spinal axons apparently do not undergo programmed cell death (apoptosis), but make connections elsewhere after withdrawal from the spinal cord, as shown in double labeling studies (O'Leary, Stanfield, and Cowan, 1981; O'Leary, 1992; Stanfield and O'Leary, 1985b). This represents one of several known examples where failure of cortical neurons to establish permanent connections at one target leads to formation of axon collaterals and to stabilization of connections on targets other than the one originally chosen. This trial and error method of initial axonal growth had been described near the turn of the nineteenth century by Santiago Ramon y Cajal (1960). It may, at least in part, account for exuberant synaptogenesis, followed by synapse elimination, during development. The ability of transient connections to be replaced by new ones may account for the relative unimportance of apoptosis (programmed cell death) during late fetal and postnatal development of the cerebral cortex. In other words, cortical neurons that fail to establish functional connections on the first try will try again and will eventually find targets that form a functional system rather than undergo apoptosis.

Hicks and D'Amato in 1970 reported a remarkable discovery when they studied the effect of hemispherectomy or of unilateral ablation of the motor cortex in infant rats. They found that animals so effected formed a large permanent ipsilateral cortico-spinal tract, which is missing in normal rats. This compensatory uncrossed motor pathway was seen in rat pups operated on between the newborn period and about postnatal day 14. The initial observations by Hicks and D'Amato have been confirmed and extended in numerous subsequent studies (Castro, 1975; Leong and Lund, 1973; D'Amato and Hicks, 1978; Sharp and Evans, 1983; Huttenlocher and Raichelson, 1989). The postnatal period during which motor cortex lesions lead to formation of the uncrossed cortico-spinal tract coincides with the period of transient ipsilateral cortico-spinal axons. It is therefore likely that some of these normally transient axons may persist and may form the new, uncrossed cortico-spinal pathway after neonatal lesions. This possibility is supported by cortical transplants of immature visual cortex neurons to the motor cortex. Such transplanted cells have been found to make permanent cortico-spinal connections (Stanfield and O'Leary, 1985a). The findings suggest that the cortical neurons of rodents remain multipotential for at least some time after birth, and may be incorporated into different functional systems. Specification of a cortical neuron for a given function may be determined by its location in the cortex and by functional demands rather than by a fixed program.

Retrograde labeling experiments, utilizing injection of HRP into spinal cord ipsilateral to the cortical lesion, have provided evidence concerning the location of the cells of origin of the new, uncrossed rodent cortico-spinal tract (Huttenlocher and Raichelson, 1989). These cells were found to be mainly in the normal motor cortex, but also in fringe areas surrounding the

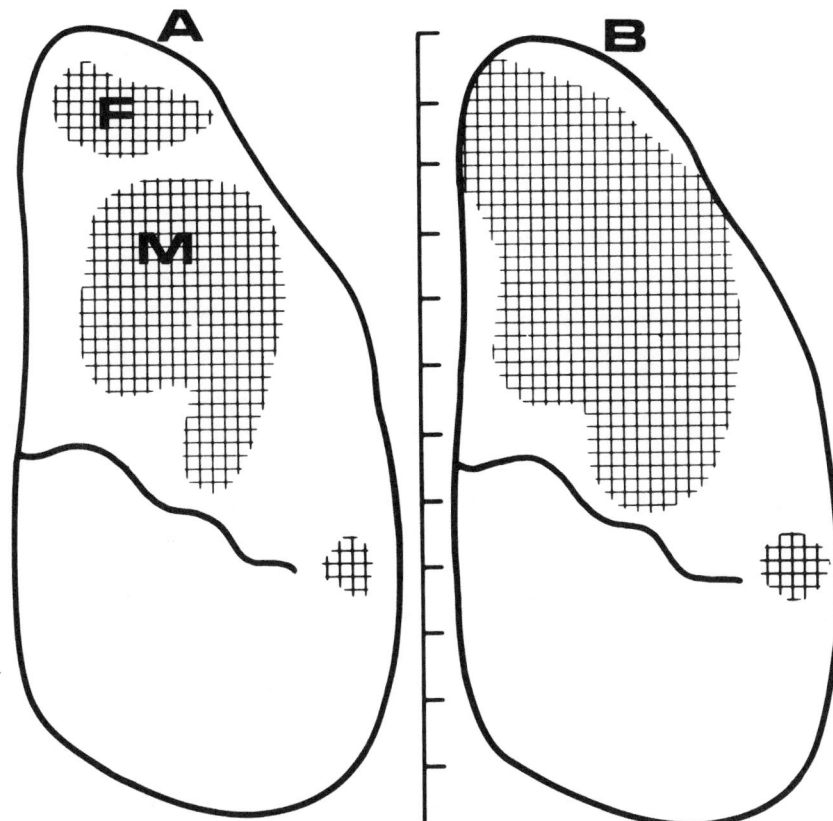

Figure 5.1. Dorsal view of a rat brain. The areas in which corticospinal neurons are found are indicated by shading. Panel A represents areas of the motor cortex in the normal rat, panel B indicates areas in which motor neurons are found after neonatal hemispherectomy. The normal rat has a prefrontal motor area (F) which is separate from the main sensorimotor cortex (M). In the single hemisphere that remains after neonatal hemispherectomy, the two motor areas are enlarged and contiguous. (Modified from P. R. Huttenlocher and Raichelson, 1989, fig. 4, p. 63, with the permission of Elsevier Science.)

motor cortex, but not as far afield as the visual cortex (Figure 5.1). Reorganization apparently was possible by recruitment of cells in the motor cortex and in immediately neighboring areas that normally do not send axons to the spinal cord in the mature state. The fact that the cells of origin of both transient axons and of axons in the uncrossed tract have been found in the same cortical areas provides a further piece of evidence in favor of the hypothesis that the uncrossed cortico-spinal tract arises from normally transient cortico-spinal axons. Other possibilities are: (1) de novo outgrowth of pyramidal tract axons from cortical neurons and (2) sprouting of an axonal branch to the ipsilateral spinal cord by neurons that also make crossed connections. The latter possibility was investigated in a double labeling study, in which two fluorescent tracers were injected into the spinal cord, one on each side, and their retrograde transport to neurons in the motor cortex was studied. No evidence for double labeling of single cortical neurons was found, which makes it unlikely that sprouting of ipsilateral axons from neurons that also have contralateral connections is an important factor (Reinoso and Castro, 1989).

The formation of a new, uncrossed cortico-spinal tract leads to some blurring of functional localization in the cerebral cortex of the rodent, with expansion of the cortical regions within which cortico-spinal neurons are seen. In that respect, the effect of cortical plasticity resembles the normal immature state, when cortical representation is diffuse rather than strictly localized. This raises the question of the utility of functional localization in the cerebral cortex. One possibility is that specification of a cortical region for one function enhances the speed and accuracy of information processing. There is, however, no direct evidence that functions of the normal crossed motor system are impaired when one cerebral hemisphere is called upon to process motor planning for both sides of the body. After neonatal hemispherectomy in the rat, the density of pyramidal tract neurons in areas outside the motor cortex is never very high, and these cells may be accommodated without major impairment of the normal functions of these areas. The number of axons in the remaining crossed (normal) cortico-spinal tract is not affected by formation of the uncrossed tract (Huttenlocher and Raichelson, 1989). Apparently, the cerebral cortex can accommodate both systems without a decrease in size of the normally occurring one. The lesion studies therefore provide no clear evidence of competition between the two sides of the spinal cord for cortico-spinal input. However, so-called crowding effects are not excluded by the data. They have been implicated in other systems, such as the developing visual cortex. They may also underlie the diffuse decrease in cognitive function, including IQ, which is seen after focal brain lesions in human infants. This problem is discussed in Chapter 7.

The uncrossed cortico-spinal tract that forms after unilateral cortical ablations appears to be of functional significance. D'Amato and Hicks (1978) already noted that nenonatal total removal of one cerebral hemisphere (hemispherectomy) resulted in rats with apparently normal motor behaviors. More recent, careful testing of such animals by Kolb and Whishaw (1989, 1998) confirms their excellent recovery and their superior motor function, compared to animals with adult lesions.

Metabolic studies confirm that the uncrossed cortico-spinal system has functional activity. Sharp and Evans (1983) studied adult rats with unilateral motor cortex lesions at the age of 1 day and compared them to animals with lesions at 30 days. The technique that was used was autoradiography after [$^{14}$C]2-deoxyglucose injection. The technique provides semi-quantitative data concerning regional cerebral metabolism of glucose. The animals with cortical ablations at the age of 1 day, but not those with lesions at the age of 30 days, had bilateral increase in uptake of the radioactive tracer in subcortical structures after electrical stimulation of the remaining motor cortex, indicating functional activity in the uncrossed motor system.

The results of neonatal hemispherectomy in rodents provide support for the Kennard Principle. They also provide information concerning a possible mechanism for neural plasticity, namely the enlistment of transient axons for the formation of permanent fiber tracts. Transient axons are known to occur in many cortical afferent and efferent systems, including the optic nerves (Rakic and Riley, 1983; Provis et al., 1985), callosal axons (Innocenti, Fiore, and Caminiti, 1977; Innocenti, 1981, 1995), and cortico-spinal axons (Stanfield and O'Leary, 1985b). They disappear fairly early in development, and plasticity that depends on their presence is unlikely to persist much beyond the neonatal period, or the fetal period in early brain developers, such as monkeys. These findings provide a further example of the large variation in the "time windows" for plasticity in different cortical systems. In part, differences in "time windows" are due to different cellular mechanisms of plasticity.

### PLASTICITY RELATED TO INPUT EFFECTS

Contrary to functional organization in the visual system, where the effects of afferent activity during development have been amply confirmed, the emergence of basic motor functions appears to be relatively unaffected by environmental input. For example, Held and Bauer (1967) raised infant monkeys in an apparatus where they were prevented from walking or from seeing their limbs. When removed from the restraints at the age of 65 days they quickly started to walk and did so almost normally after 1 week. No permanent deficits were seen. Hein and Held (1967) found that 4-week-old kittens reared without sight of limbs or torso had a normal placing reaction (extension of the forelimb) on approach of a horizontal surface. This simple, visually guided motor response appears to be "innate," or independent of prior experience. In contrast, more complex visually guided behavior, such as reaching and avoidance of a discontinuity in a surface (a visual cliff), requires experience and practice, including viewing of the limb, and develops slowly after removal of restraints. Most likely, the basic programs for walking and for visually guided placing of a forelimb are intrinsic, determined by genetic programs. The ability to walk exists in every mammal with a normal central nervous system. It emerges without specific training during a rather narrow "time window," consistent with a fixed developmental program. However, not all voluntary motor functions are genetically determined. While the animals in the Held and Hein experiment were able to walk almost immediately after release from their restraints, they did show transient deficits in visually guided motor behaviors.

Practice clearly is important for the development of superior motor abilities. There also is evidence for a "time window," a period during development when practice of highly skilled motor tasks is especially effective. Gymnasts must start their training in childhood, and tend to reach the zenith of their skill by adolescence. The observations on motor development, like those related to the development of basic visual functions (Chapter 4), suggest that experience is necessary for the emergence of complex, often elective, functions, while more basic, obligatory ones, such as the placing reaction, appear to develop largely if not completely independently of experience.

Animal experiments have shown that the anatomy of the developing motor system is influenced by functional demands. Pysh and Weiss (1979) found that the cerebellar Purkinje cells of mice had significantly larger dendritic trees when the animals were allowed to exercise from the age of 18 days (weaning) to the age of 35 days. The control group consisted of littermates raised in crowded cages without exercise equipment. The effects of exercise on cortical anatomy have been found even in adult rodents (Greenough, Black, and Wallace, 1987; see also Chapter 8).

Knowledge concerning plasticity in the motor cortex in humans derives primarily from the study of patients with early-acquired focal brain lesions. Most of the information is from cases of stroke, either in utero or in the perinatal period. In addition, there is information concerning voluntary motor functions in patients with hemispherectomy. These in some respect provide definitive information about the ability of the single remaining motor cortex to mediate ipsilateral as well as the normal contralateral motor functions. A problem that is encountered in the interpretation of data from hemispherectomy cases is the fact that there are two lesions, acquired at different ages. The first may be a unilateral pre- or perinatal lesion or a lesion caused by a progressive inflammatory disease during childhood (chronic focal encephalitis or Rasmussen disease). This initial lesion subsequently caused intractable seizures, which in turn led to the decision to perform hemispherectomy, often several years after the date of the initial lesion. In the meantime, both the seizure activity and the occurrence of a focal brain lesion early in life are likely to have resulted in reorganization of the cerebral cortex. Superimposed on this is the reorganization caused by hemispherectomy. Nevertheless, hemispherectomy cases provide important information, especially selected ones with early hemispherectomy and without complicating conditions, such as prolonged, intractable seizures. Such cases are likely to provide information about the upper limits of functional reorganization in a person with a single motor cortex.

## Sparing of Facial Movements and of Gait

Interesting effects of age at the time of stroke on motor function have been reported relative to facial movements. Children with a history of pre- or perinatal strokes usually do not have facial weakness. The cosmetic defect of facial palsy and the drooling of saliva and difficulties in eating that complicate stroke in the older child or adult are not seen in the early lesion cases (Lenn and Freinkel, 1989). This lesion-related plasticity in the cortico-bulbar system disappears at or near to term.

Lower extremity functions also are remarkably good in children with early lesions, especially as these relate to walking. These children, even those with hemispherectomy, have relatively mild gait abnormalities (Ueki, 1966). There is usually only a mild hemiparetic limp and mild spasticity. The age at which walking occurs is little delayed, if at all. The upper age limit for preservation of good gait patterns after motor cortex lesions has not been determined.

Sparing of facial movements and of gait in infantile hemiplegia is likely to relate to axonal growth and retraction rather than to synaptic mechanisms. This is suggested—at least for facial sparing—by the early disappearance of plasticity, in the immediate postnatal period, long before synapse elimination occurs in the motor cortex. Motor plasticity may relate to the formation of transient cortico-bulbar connections from motor cortex to the ipsilateral medullary gray matter, which are withdrawn during development. Such transient axons are very prominent in the fetal and neonatal rat (see above). Transient cortico-bulbar and cortico-spinal axons, including uncrossed ones, are likely to occur in humans as well. They may be utilized for functions such as facial movement after perinatal motor cortex injury. Incorporation into a functioning system may stabilize these connections. It is of interest that the part of the facial nucleus that innervates the forehead muscles has little direct cortical input from either side. Its cortical input is indirect and bilateral, via at least one synapse in the brain stem reticular formation (Jenny and Saper, 1987). This arrangement may facilitate bilateral cortico-bulbar interactions. A unilateral cortical lesion spares the contralateral upper facial movements, even when the lesion occurs in an adult.

### LIMITED RECOVERY OF VOLUNTARY HAND AND FINGER MOVEMENTS

Recovery of hand functions after neonatal motor cortex lesions is much more limited than is that of facial movements and of gait. Ueki (1966) found that most patients with hemispherectomy had no useful voluntary finger movements. The same is true for children with large perinatal strokes involving the motor hand area. What these children frequently have is mirror movements of the paretic hand and arm (Nass, 1985). Mirror movements are involuntary participations of a limb that more or less reproduce voluntary movements on the other side. For example, voluntary apposition of thumb and forefinger in one hand will be accompanied by involuntary simultaneous reproduction of these movements in the other hand. Mirror movements do not provide useful motor function, and in children who have some voluntary hand movements they may actually interfere with motor activities.

Normally, there is a small uncrossed cortico-spinal pathway in adult humans. Mirror movements are thought to be due to activation of this uncrossed pathway. The utilization of uncrossed cortico-spinal axons for mirror movements is supported by the observation that these movements are very prominent in some patients with Klippel-Feil syndrome, a syndrome of congenital malformations in the region of the cervical spine. In this syndrome, there may be failure of the normal decussation (crossing) of the pyramidal tract fibers in the lower medulla and upper spinal cord, resulting in a large uncrossed component of the cortico-spinal system (Gunderson and Solitaire, 1968). In patients with infantile hemiplegia, there is evidence from fMRI studies that uncrossed connections between motor cortex ipsilateral to the paralysis and the spinal cord are increased (see below).

Why is recovery of hand movements after perinatal motor cortex lesions so much poorer than that of facial and leg movements? One important difference between movements of the face and legs and those of the hands and fingers lies in the type of motor control that is involved. Facial movements are normally bilaterally symmetric and synchronous, as in smiling. There is only a small step from mirror movements, which are known to be mediated by the ipsilateral motor cortex, to the facial movements of smiling and talking, which are by and large bilaterally symmetric. Relocation of this type of simple bilateral function in the ipsilateral cortex may be relatively simple. The movements of walking also are mirror movements, which alternate in phase in a precise manner. The movements of one leg in walking accurately determine the movements required of the other leg. This reproduction of an identical movement again is a fairly simple task that may be reorganized in the ipsilateral cortex or in subcortical areas. Hand and finger movements, on the other hand, are much more complex. They are asymmetric and involve planning of complex sequences of muscular contraction and relaxation. They require information processing outside of primary motor areas. Evarts, Shinoda, and Wise (1984) reported that in monkeys carrying out skilled motor acts there always is activation of neurons in the postcentral gyrus prior to activation of the motor cortex proper. Supplementary motor cortex, located anterior to the hand area in the precentral gyrus, also becomes activated in relation to finger movements. The complex intracortical connections necessary for skilled movements may not be reproducible in the ipsilateral cortex.

Another factor in poor recovery may be competition between the right and left motor hand area for dominance. While there are differences in the degree of lateralization of hand functions, one hand emerges as the dominant one in most humans. The development of hand dominance is at least partially environmentally determined. Children with a tendency to left-handedness can be trained to use the right hand preferentially. This appears to account for the virtual absence of left-handedness in some societies (Harris, 2000). The child with infantile hemiparesis has a strong bias toward using the unaffected hand. He or she therefore has, in a manner of speaking, continuous and intensive training for preferential use of the unaffected hand. The infant with weakness in one hand will soon neglect that hand in favor of the normal side when carrying out skilled motor tasks.

The question arises whether this neglect of the paretic hand and preferential use of the normal hand could be prevented by early training. Some efforts have been made to force infants with hemiparesis to practice use of the paretic hand by restraining the normal hand (Taub and Crago, 1995). However, a comprehensive program of prolonged encouragement to use the paretic hand, started early in infancy, has not been critically evaluated. The experience with patching of the dominant eye in the infant with strabismus, discussed in Chapter 3, suggests the possibility that such a program, initiated prior to the end of the critical period, could improve paretic hand function. The limits of such a critical period in the motor hand area are not well established. The sparing of facial movements after early strokes suggests that these movements are preserved only when the lesion occurs in the prenatal or perinatal period. The critical period in the pyramidal motor system therefore may end prior to the time when remedial training could be instituted. However, there may be differences between the "time window" for spontaneous recovery and that for improvement related to training (see also Chapter 7 for the effects of practice on performance in childhood). Careful study of intermittent limb restraint and/or of intensive practicing of paretic hand use starting in early infancy appears indicated, especially in view of the results recently obtained with restraint of the normal hand and intensive practice of the impaired one in adult stroke patients (Taub et al., 1993). One must, however, keep in mind that negative effects of unilateral limb restraint in infants with hemiparesis may occur, owing to the inhibition of normal manual exploration when the competent hand is restrained. Such a program should therefore be critically evaluated in controlled studies before widespread use is considered.

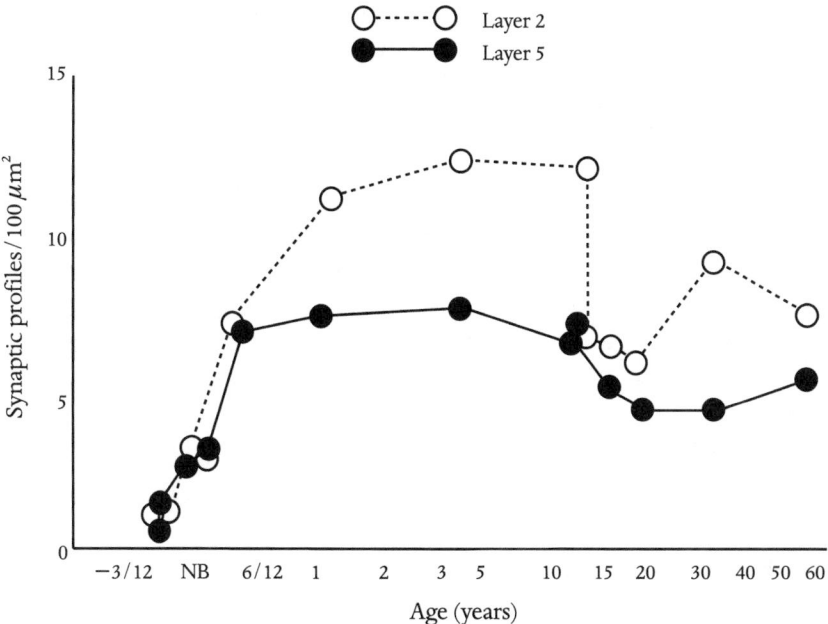

Figure 5.2. Synaptic density in human motor cortex (left-hand area) as a function of age (solid line = layer 5; dashed line = layer 2).

Poor recovery of voluntary hand functions after perinatal injury to the motor cortex may be related to special aspects in the development and organization of this cortical area. The developmental anatomy of the motor cortex differs from that of other cortical regions. An obvious anatomical difference is the occurrence of the giant Betz cells in layer 5, whose axons form the cortico-spinal tracts. At all ages studied, layer 5 of the motor cortex has the lowest synaptic density of any cortical region so far examined. Maximum synaptic density is about 60 percent that of layer 5 in other cortical regions. The adult synaptic density also is low, about 60 percent of layer 5 synaptic density in other cortical regions (Figure 5.2). The difference in synaptic density between layer 5 in the motor cortex and in other cortical areas most likely relates to the fact that layer 5 of the motor cortex has relatively little neuropil in which synapses may form, since a major part of cortical volume is occupied by the cell bodies and initial axonal and dendritic segments of very large pyramidal neurons, including the Betz cells, that are devoid of synapses. The presence of these large cells makes it possible to transmit information over remarkably long distances from motor cortex to spinal cord, without the delay and modification of the message that would result from interposed synapses. In layer 5 of the motor cortex, the type of plasticity that depends on the availability of large numbers of unspecified synapses may be diminished in favor of a relatively fixed, very efficient, rapidly conducting system that allows for remarkable speed and precision of voluntary movements. Maximum synaptic density in the motor cortex is reached early, at the age of 5–6 months, at much the same time as in the visual and auditory cortex. Overproduction of synapses during infancy is less marked than in other cortical areas, consistent with limited plasticity. Synaptic density decreases to adult values during late childhood and early adolescence, much as in most other cortical areas (Figure 5.2). While the data on synaptogenesis in the motor cortex show differences from those on synaptogenesis in other cortical areas, the functional significance of these differences, and whether they underlie the relatively low ability of the motor cortex to reorganize in response to focal lesions, are as yet speculative. The limited evidence derived from animal studies suggests that low plasticity in motor recovery may relate to the necessity for formation of a functioning uncrossed cortico-spinal system, which cannot be accomplished after the neonatal period.

### FUNCTIONAL MRI STUDY OF REORGANIZATION OF THE MOTOR CORTEX AFTER PERINATAL INJURY

Hypotheses concerning the mechanism of functional plasticity can now be tested noninvasively in vivo by functional MRI. Cao, Vikingstad, Huttenlocher, and colleagues (1994) studied a group of patients with infantile hemiparesis by this technique, in order to determine whether reorganization similar to that in the hemispherectomized rat pup could be demonstrated in humans. The patients had histories of either pre- or perinatal stroke or of developmental malformation of one cerebral hemisphere. The fMRI studies were carried out many years later, during adolescence or young adulthood. The subjects were of low normal to normal intelligence. They were selected on the basis of strictly unilateral brain lesions and the presence of hemiparesis, but preservation of the ability to voluntarily appose thumb and forefinger on the paretic side. The latter constraint was necessary since the method depends on performance of a simple motor act (such as apposition of thumb and forefingers) during the scan. Results obtained while the subject was at rest were compared to those collected while the subject was engaged in repeated thumb-forefinger apposition. This study showed evidence of reorganization of the motor cortex function to the side ipsilateral to the hemiparetic side in all subjects. The area of brain on the anatomically normal side that was activated during voluntary motor activity was larger than that in controls, a result similar to the finding in rodents with neonatal hemispherectomy (see Figure 5.1). In the human, the additional area that is activated is located posterior to the motor cortex, in the area of the postcentral gyrus. This cortical region normally has many connections with the motor cortex and is thought to be involved in motor planning (Evarts, Shinoda, and Weiss, 1984).

The demonstration of increased cerebral blood flow ipsilateral to the side of voluntary motor activity suggests that the functional importance of the uncrossed cortico-spinal pathway increases after neonatal damage to one hemisphere. It is not known whether this is due to persistence of uncrossed cortico-spinal axons that would otherwise undergo normal retraction, or whether there is facilitation of synaptic circuits that are normally inactive. In normal controls, there was a small ipsilateral area of activation with left (nondominant) finger movements only. Bilateral activation of motor cortex during motor task performance in the nondominant hand of normal subjects has been observed in other fMRI studies as well (Cao et al., 1993). This finding suggests that the uncrossed cortico-spinal pathway maintains some functional activity normally.

Our hemiparetic subjects had better hand function than that described after total hemispherectomy or after hemidecortication. Almost all of them had some residual activation in the damaged hemisphere, related to movement of the paretic fingers. It therefore is likely that their finger movements depended on activation of residual motor cortex both on the damaged side and on the ipsilateral, anatomically normal side. It is important to consider that some functional activity may persist in even a severely damaged cerebral hemisphere. Surgical removal of such a hemisphere may lead to an increase in functional deficit, especially in the paretic hand. Careful mapping of cortical motor areas prior to surgery therefore is essential. The functional localization of cortical motor areas in humans can now be carried out relatively noninvasively by fMRI and by transcranial magnetic stimulation (Pascual-Leone et al., 1994).

The recovery of motor function after adult stroke varies widely, depending on the nature of the lesion. The initially observed motor deficit often improves over a period of several days or weeks. This is probably in part related to recovery of function in neuronal circuits near the edge of the infarct or area of tissue necrosis. Activity in nearby undamaged tissue may be temporarily disrupted by loss of input from neighboring damaged brain areas, and by such factors as edema and the release of neurotoxic substances from damaged neurons. Much work has focused on excitotoxicity due to the release of glutamate, the major excitatory neurotransmitter substance in the neocortex (Rothman and Olney, 1986). The contribution of these reversible factors to the initial functional deficits is not predictable in any given case. For any comparisons between effects of neonatal and adult strokes, it therefore is necessary to use data collected after the deficit has stabilized. The importance of this has been pointed out already in the monkey studies of Kennard and Passingham and their colleagues, discussed earlier in this chapter. There are at present no studies in humans comparing the effects of similar-sized lesions in similar locations sustained at different ages. However, certain differences between perinatal and adult lesions are clearly present. These include sparing of facial movements in prenatal strokes, and the development of good gait patterns in infants with lesions, even after hemispherectomy. These children have a hemiparetic limp, but have little functional impairment and often become quite competent in team sports. Total unilateral destruction of cortex and of the subcortical white matter in the adult, on the other hand, leads to severe gait disturbance (Collen, Wade, and Bradshaw, 1990). Differences in hand function between adult and neonatal strokes are less clear, and would require careful study of cases matched for lesion size and location, and time from occurrence of the lesion to study.

It is known that some degree of reorganization of the pyramidal motor system occurs after partial lesions in the adult. PET studies of adult patients with a history of striatocapsular stroke have shown ipsilateral increase in metabolism of sensorimotor cortex, similar to the fMRI findings in patients with perinatal stroke (Weiller et al., 1992). Seitz and colleagues (1998) reported a group of adults with marked recovery of hand function after middle cerebral artery territory infarction. TMS, MRI, fMRI, PET, and evoked potential studies were carried out. The lesions involved the cerebral convexity along the Rolandic fissure from the Sylvian fissure up to the hand area, but spared the dorsal precentral gyrus, basal ganglia, and middle and posterior portions of the internal capsule. During individual finger movements of the paretic hand there was increase in cerebral blood flow in bilateral premotor areas. No such increase was seen in the sensorimotor cortex (pre- and postcentral gyri) on either side. The results suggest that in these patients with partial lesions the motor cortex was able to reorganize in supplementary motor areas located anterior to the precentral gyrus. Utilization of subcortical structures may also be of importance for recovery from adult strokes. Subcortical gray matter areas important for motor recovery include the basal ganglia, the thalamus, and the cerebellum (Seitz et al., 1999).

What emerges is that in the adult, recovery depends to a considerable extent on lesion location. Reorganization in undamaged brain regions that normally have some motor function appears to be an important factor. In that regard, it is important to stress that motor representation in the cerebral cortex is normally complex, involving supplementary motor areas (M2, anterior to the rostral portion of primary motor cortex, M1). Even in M1, cortical localization is not as precise as is implied by the classic "homunculus." Joint movements rather than individual muscles are represented, and a given joint movement activates multiple patches of cortex (H. J. Gould, Kusik, Pons, and Kaas, 1986). The normal activation of multiple patches of tissue during performance of a simple motor task is well demonstrated by fMRI.

Even in a system that is apparently largely specified prior to birth, some late reorganization of functional localization can occur. The expansion of the cortical representation of finger movements in the left hand of violinists (Elbert et al., 1995), discussed in Chapter 7, is a good example.

### Summary

Plasticity in the motor cortex differs from that in other cortical regions, in that there is limited ability of other groups of neurons to "take over" functions that have been lost owing to focal damage to the brain. This appears to relate to the early specification of the motor cortex, including the formation of a cortical map, the well-known "homunculus." The efferent system from cortex to spinal cord is dependent on the giant Betz cells in layer 5, which apparently have limited capacity for reorganization after focal motor cortex injury. Whatever lesion-related plasticity occurs in this system appears to be due to increased activity in the uncrossed cortico-spinal pathway. This may occur either through stabilization of normally transient ipsilateral cortico-spinal axons or due to sprouting of new ipsilateral connections. Animal studies favor the former alternative. Plasticity related to stabilization of transient efferent connections ends early in development, around the time of birth in the human. An unresolved question is whether disuse of the paretic hand due to preferential use of the normal hand may be a factor in the severity of the functional deficit. This question needs to be resolved by carefully designed studies of intermittent restraint of the normal hand in infants or young children with hemiparesis.

# 6

## PLASTICITY IN THE DEVELOPMENT OF LANGUAGE

> Within the human species, we find little evidence for a circumscribed
> language organ. The whole brain participates in language, although
> the pattern of participation that we see varies, depending on the task
> at hand, and some regions are clearly more important than others.
>
> Elizabeth Bates, 1993

Language functions are uniquely human. No animal models are available, except for the higher apes, which appear to be able to learn a simple receptive language, including some rudiments of grammar (Savage-Rumbaugh et al., 1993). There also are some interesting parallels between human language and bird song (see below). However, almost all of our knowledge is derived from human studies and observations, both in normal subjects at various ages and in patients with focal brain lesions. Nevertheless, a large body of information is available.

In no area of cortical function is age-related plasticity as evident as in language. This is especially so in relation to the development of language after unilateral perinatal brain damage. Focal brain injury that destroys the classic language areas of the brain leads to aphasia in adults, but not in infants and young children. This difference was known already in the nineteenth century. Sigmund Freud discusses it in his classic 1897 monograph on cerebral palsy. Basser (1962) provided the first careful neuropsychologic assessments of a group of patients with large early-onset unilateral cortical lesions, treated with hemispherectomy later in life, usually for control of intractable seizures. A large literature on the effects of early (usually perinatal) unilateral brain lesions has appeared since then. The results and interpretations have not been without controversy.

Marked environmental influences on language development have also been described in the normal infant and child. The acquisition of vocabulary and grammar shows marked dependence on environmental input. There is increasing evidence that impoverished language skills may be related to decreased early language input.

The young child has long been known to have an edge over the adult when it comes to learning a second language. This appears to be relatively effortless in the child, and—perhaps more important—is more perfect than in the adult. The young child learns a second language without accent and with normal grammar, a feat that is not possible for most adults. Interesting recent evidence indicates that the ability of the child to hear subtle differences in speech sounds depends on the language or languages she or he was exposed to early in life. Unfamiliar sound discriminations, which may be important for flawless learning of a second language, may no longer be possible after a certain age.

### Lesion-Related Plasticity in Language Functions

There appears to be universal agreement that language functions after focal injury such as stroke in infancy are remarkably different from those after stroke in the adult. In adults, damage to classic language areas of the brain, usually in the left cerebral hemisphere, produces more or less specific syndromes of aphasia, which either are permanent or improve slowly and usually incompletely. Widespread damage to the dominant (usually left) hemisphere causes global aphasia, in which both receptive and expressive language functions are lost. Damage to Wernicke's area (Figure 6.1) in the left perisylvian (surrounding the Sylvian fissure) temporal lobe is associated with decreased language comprehension, often with very fluent "speech" that approximates normal speech to some extent, but is incomprehensible and often contains nonexistent words or words similar in sound but different in meaning from the appropriate words ("paraphasias"). This syndrome, referred to as "Wernicke aphasia," is not seen in similar lesions incurred during infancy or in childhood up to the age of 8 or 9 years.

A different form of aphasia is seen in large left frontal lesions in the adult that include Broca's area (Figure 6.1), but almost always extend beyond it,

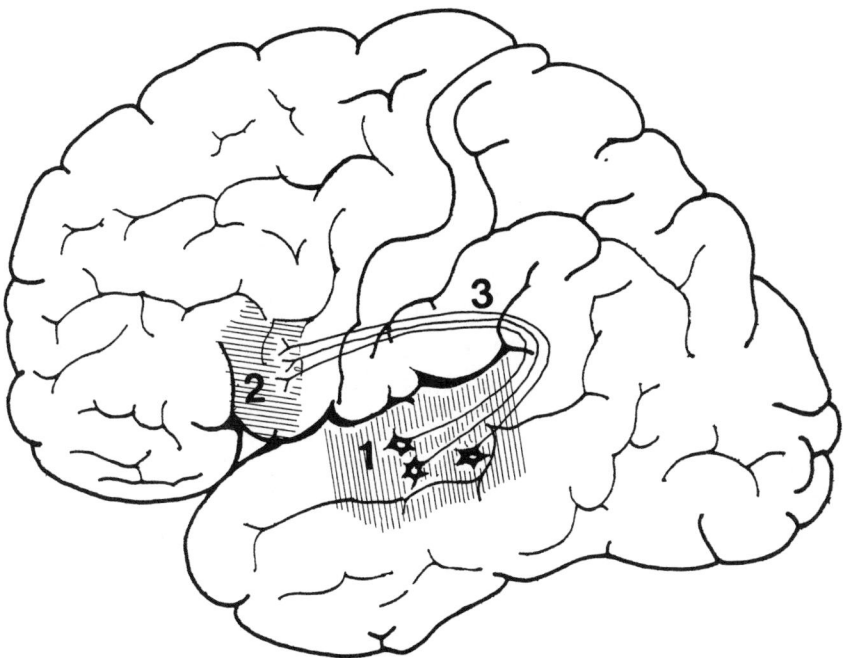

Figure 6.1. Left lateral view of the human brain. Wernicke's area (1) and Broca's area (2) are shaded. The arcuate fasciculus (3) connects these two cortical regions.

often anteriorly. Language comprehension is preserved, but language production is markedly impaired, ranging from complete muteness to dysfluent, simple speech. Something resembling Broca's aphasia clinically, but not in the location of the lesion, is seen acutely, in a very transient form, in early childhood focal brain lesions. A third type of adult aphasia, referred to as conduction aphasia, is due to a lesion in the arcuate fasciculus (Figure 6.1), the large white matter tract that connects Wernicke's area with Broca's. Language comprehension is preserved, but speech production is abnormal, with many paraphasic errors. Again, this syndrome does not occur in childhood.

Acute focal brain lesions in infants prior to the appearance of language are not associated with marked aphasic disorders later in life. Most of these lesions appear to be strokes (due to occlusion of a cerebral artery) that occur in the perinatal period. The age of onset of speech tends to be somewhat delayed in such children. Receptive language develops at a more normal rate. In infants with focal brain lesions incurred prior to the age of 6 months, speech delay is seen following right hemisphere as well as left hemisphere lesions. Lesions in Wernicke's area (left posterior superior temporal cortex) in the infant are associated with delayed language production, not with the defects in language comprehension seen in the adult (Marchman, Miller, and Bates, 1991; Thal et al., 1991; Bates, 1993). Subsequent language development appears to progress normally or nearly so. It is here where a point of disagreement appears. Several reports, especially early ones, find that there is no difference in language skills between left and right hemisphere damaged children (Freud, 1897; Basser, 1962). Basser studied a large population of over 100 patients with early onset focal brain lesions, including about 30 with hemispherectomy (total removal of a cerebral hemisphere). Both the subjects with right- and with left-sided perinatal lesions showed somewhat delayed onset of speech, as had been reported earlier by Freud. Elizabeth Bates and her coworkers also found that most children with early focal unilateral brain lesions eventually achieved language functions that were within the normal range (Bates, 1993). However, several investigators have reported language deficits in children with a history of early left hemisphere damage, when they are compared to sibling controls and to children with early right hemisphere injury (Dennis and Whitaker, 1976; Rankin, Aram, and Horwitz, 1981; Vargha-Khadem et al., 1985; Aram et al., 1985, 1986, 1987; Witelson, 1987; Eisele and Aram, 1994, 1995). Vargha-Khadem and colleagues (1991, 1992) studied patients with hemispherectomy, and found language deficits in all subjects with left hemidecortication. Language development also was abnormal in right hemispherectomy patients whose initial brain insult had occurred in early childhood. The observed deficits have been primarily in the more complex aspects of language, especially grammar or syntax comprehension, and in the fluency of speech. The lexicon (vocabulary) is affected less. The choice of an appropriate comparison group is important for this type of study. Decreased performance in relation to sibling controls is not surprising, since early unilateral brain injury is associated with slightly below normal overall IQ (see Chapter 7). Right hemisphere injured children may also be an imperfect control group, unless they are matched for lesion size and location, since—for unexplained reasons—perinatal right hemisphere lesions are on the average smaller than those on the left and since lesion size and IQ are negatively correlated (Levine et al., 1987). Nevertheless, most of the available data suggest that preservation of the left cerebral hemisphere, including Wernicke's area, imparts a slight advantage as far as language functions are concerned, even in the case of early infancy lesions, and even when lesion size is held constant, as is the case in subjects with hemispherectomy. At the same time, it needs to be stressed that language deficits in subjects with early damage to the classical speech areas are subtle, of limited functional significance for the patient, and usually apparent only after detailed psycholinguistic testing, very different from the outcome in adult dominant hemisphere stroke.

Up to now I have considered outcomes in children with focal brain lesions prior to the onset of speech, usually in the pre-and perinatal periods. The results of strokes that occur in children who are past the age of acquisition of language also differ from those in adults. Freud (1897) reported a transient decrease in language production without any deficit in comprehension in patients with childhood strokes, rather than adult types of aphasia. This outcome has been described vividly in case reports collected by Basser (1962). Muteness or sparsity of speech is followed by rapid return of seemingly normal language functions, usually over a period of days or weeks (Basser, 1962; Woods and Teuber, 1978). Transient loss of expressive language may occur in lesions involving Wernicke's area, Broca's area, or both. The occurrence of transient muteness after right hemisphere as well as after left hemisphere lesions in young children, up to the age of 2 or 3 years, suggests that the right hemisphere is important for language functions in the young child. Basser (1962) reported aphasia in nearly 50 percent (7 out of 15 cases) of right hemisphere lesions in children past the onset of speech at the time of the stroke. However, modern imaging modalities were not available at the time of his study, and some of these cases may have had bilateral lesions. The incidence of aphasia after right hemisphere lesions in childhood was found to be lower by Woods and Teuber (1978) in a carefully studied series of cases.

The cerebellum also appears to have some influence on language functions in children. Focal lesions in the right cerebellar hemisphere in early childhood have been reported to cause transient muteness similar to what is seen with neocortical lesions (Herb and Thyen, 1992; van Dongen et al., 1994). The right cerebellar hemisphere apparently continues to play some role in language processing in the adult (Silveri, Leggio, and Molinary, 1994).

The adult pattern of aphasia after focal damage to the dominant cerebral hemisphere emerges gradually in late childhood. While 8 years has been stated as a critical age beyond which recovery is no longer nearly complete (Woods and Teuber, 1978), there appears to be considerable individual variability, and there is no sudden age-related change from remarkable recovery to persistent, severe language deficit with lesions of comparable size and location (Woods and Carey, 1979). Thus recovery in an adolescent is likely to be better than that in an adult, and even during the adult years, the outlook may be somewhat better in the younger age groups. Within the adolescent and adult age ranges, data are needed concerning the effect of the age of occurrence of stroke and gender on outcome.

### The Anatomical Substrate of Lesion-Related Plasticity in Language Functions

What is the underlying anatomical substrate of the remarkable recovery of language functions in the young? Reorganization of language areas in the damaged left hemisphere, de novo organization of language in the normal right hemisphere, and initial bilateral representation of language in both hemispheres have all been suggested as possible anatomic substrates for preservation or recovery of language functions in children. One may also ask whether language representation in the right hemisphere, if it occurs, is in analogous areas, namely in the right posterior superior and middle temporal gyrus (the cortical region referred to as Wernicke's area in the dominant hemisphere) and in the right posterior lateral frontal cortex (the region known as Broca's area on the left). Answers to these questions depend on careful study of humans with early-acquired focal cortical lesions, since no animal models of aphasia are available, and on the use of modern methods, including fMRI, TMS, PET, and event-related evoked potentials. At this point, the relevant data are rather sparse.

We know that in at least some patients with focal left hemisphere lesions language processing is taken over by the intact right hemisphere. Information concerning this is derived from the study of patients who have had hemispherectomy on the dominant side (explicitly removal or disconnection of the entire left hemisphere). Hemispherectomy has been carried out in several groups of children, particularly (1) children with intractable seizures due to extensive malformation of one cerebral hemisphere, (2) children with intractable seizures due to early-acquired unilateral brain damage, often stroke in the perinatal period, and (3) children with a progressive inflammatory disease of one hemisphere, so-called chronic focal encephalitis (Rasmussen disease). This disease, which is as yet of unknown cause, leads to destruction of the involved cerebral hemisphere and continuous focal seizure activity (epilepsia partialis continua) (Andermann et al., 1991). In patients who have removal of the dominant (left) cerebral hemisphere, language functions are relatively well preserved provided the initial damage to the hemisphere occurred prior to the age of 8 or 9 years. This observation does not answer the question whether language is bilaterally represented in the normal child, or whether it becomes organized de novo in the nondominant hemisphere following the injury. Nor does it provide information concerning the areas of representation of language functions in the remaining hemisphere.

Data concerning the question of bilateral representation of language in the infant have been derived from clinical observations and from evoked potential and imaging studies. Of interest is the observation that damage to the right as well as to the left cerebral hemisphere may lead to the syndrome of transient mutism in young children. Nondominant hemisphere lesions in the adult do not produce aphasia. This suggests a difference between the child's brain and the adult's brain as far as organization of language is concerned. It is consistent with initial bilateral representation and maturational loss of representation in the nondominant hemisphere.

The claim that transient language deficits occur in children with acute nondominant hemisphere damage is not accepted by all. For example, Woods and Teuber conclude that some such cases may have had right hemisphere dominance from the start, while others may have had undiagnosed bilateral lesions (Woods and Teuber, 1978). However, in view of what we now know about the low frequency of right hemisphere dominance for speech, even in left-handed people, it is unlikely that such cases could explain the fairly frequent occurrence of speech involvement in right hemisphere childhood strokes.

Some aspects of language are known to be normally represented in the nondominant hemisphere in the adult brain. Subjects with nondominant hemisphere lesions of adult onset often have aprosody: monotonous, uninflected speech that lacks the musical aspects of language. This has also been reported in a childhood syndrome of "nonverbal learning disability," presumably secondary to dysfunction of the nondominant hemisphere (Weintraub and Mesulam, 1983; Voeller, 1986). The dominant hemisphere of such children does not appear to be able to compensate for the loss of this right hemisphere function. There may be less plasticity in prosody than in other aspects of language. This is consistent with other findings that suggest a generally lower capacity for reorganization of nondominant hemisphere functions, including—besides prosody—spatial and constructional abilities (see Chapter 7).

There is some suggestion that the right hemisphere may participate in some language functions other than prosody in the adult. This has been suggested to be so for extraordinarily complex language tasks, including, for example, reading comprehension of scientific writing and composition of scientific manuscripts. It is usually difficult to test whether such a function has been impaired in a nondominant hemisphere stroke, since baseline testing of extraordinary skills prior to stroke is usually not available. The anatomist Alf Brodal provided a vivid description of subtle, language-related deficits as he observed them in his own case (Brodal, 1973). He had what appears to have been an embolic stroke, affecting the right internal capsule. Clinical testing revealed an apparent pure motor hemiplegia on the left. There was no aphasia. Yet Brodal noted that his ability to extract the content of a scientific article was impaired. He also compared the composition of the beginning of a letter, which he had started prior to the stroke, with that of the end of the same letter, completed after the stroke, and found the second part lacking in sophistication. There also appeared to be a subtle effect on handwriting. The conclusion Brodal derived from his case, that the nondominant hemisphere may be involved in complex language processing, is supported by some recent PET and fMRI studies, which show right as well as left cortical activation during performance of more difficult language tasks (see below). This appears to be an example of a general principle of functional localization in the cerebral cortex, namely that the activated area becomes larger, and information processing becomes more diffusely distributed in the cerebral cortex as the difficulty of the task increases. This may at least in part explain diffuse activation of the brain related to language tasks in infancy. What may be a simple, routine task for the older child or adult may be difficult for the infant, hence the entire cortex may participate in processing of the task (see the quote by Elizabeth Bates at the beginning of the chapter). As the task becomes more routine, only small regions—the classic language areas—may be needed for processing. Such a developmental progression would predict that processing of any given language task would become more localized earlier in early language learners. Evidence in favor of this comes from recent evoked potential and brain imaging studies (see below). Circuits in the right cerebral hemisphere used for the processing of simple language functions in the infant may gradually become less effective and may disappear because of disuse. This would not be the case for language tasks that are difficult for the adult, such as scientific writing, the processing of which involves large bilateral cortical regions throughout the life span.

## Language Development and Synaptogenesis

The onset of receptive and expressive language occurs at about the age at which synaptic density in the corresponding language areas (Wernicke and Broca) approaches the maximum, that is to say at about the age of 1 year. Synaptogenesis in Wernicke's area slightly precedes that in Broca's area, which perhaps reflects the earlier onset of receptive versus expressive language (Figure 2.7).

Synapse elimination in language areas occurs normally in late childhood and is complete by mid-adolescence (Figure 2.7). This time course corresponds closely to the age at which plasticity for language decreases, specifically when recovery of aphasia after a stroke diminishes to what is seen in the adult, and when second language learning becomes imperfect. By late adolescence, the right hemisphere has lost the ability to process most language functions. In general, the emergence of competence in the basic functions of a cortical area appears to coincide with the age when cortical synapses show rapid proliferation. The period during which synaptic density in language areas is above adult levels coincides with a period of increased plasticity, as has already been noted for the visual cortex, in relation to strabismic amblyopia (Chapter 4).

## Evidence from Neurophysiologic and Neuroimaging Studies

Interesting data concerning the question of early childhood bilateral representation of language have been derived from the study of cortical evoked potentials. Mills, Coffey-Corina, and Neville (1993, 1997) studied event-related potentials (ERPs) in response to presentation of single words in 13- to 20-month-old infants. At the ages of 13–17 months, ERP differences between comprehended and unknown words were widely distributed over both cerebral hemispheres. By the age of 20 months there was localization of the responses to comprehended words in the left temporal and parietal areas. Infants with precocious language development showed earlier restriction of responses to the dominant hemisphere. This study provides confirmation of the principle that tasks that are difficult engage large cortical areas, while similar but easier tasks are processed in more restricted cortical regions. Cerebral energy expenditure and glucose consumption would be expected to decrease with decrease in the difficulty of the task. This has indeed been found, when PET was used to compare the cerebral glucose consumption of slow and fast learners of a computer game during the task (Haier et al., 1992). On first exposure to the game both fast and slow learners had high cerebral glucose consumption. As the game was learned, fast learners had restriction of the activated areas and a decrease in glucose requirement. These changes were less marked in slow learners.

Activation of larger cortical regions during processing of more difficult tasks has also been noted within subjects, when task complexity was progressively increased. Just and colleagues (1996) used fMRI to study cerebral blood flow changes in response to a sentence comprehension task in adult subjects. Brain activation during visual presentation of simple sentences was confined to the classic language areas in the left cerebral hemisphere, while more complex sentences produced bilateral activation. These findings suggest that the language areas of the cerebral cortex are not fixed, but depend on the task that is provided plus the language competence of the subject. Decrease in task complexity and increase in competence are associated with a decrease in the amount of tissue that is involved in the processing of a task.

Language learning in the young child is a herculean task. An early "word spurt" has been described between the ages of 14 and 22 months in most children, although word acquisition appears to be more gradual in some (Goldfield and Reznick, 1990). During this early word spurt, the child may learn up to an average of 3 new words per day, mainly object words (Fenson, Dale, et al., 1994). Vocabulary growth is even more rapid later in childhood (Anglin, 1993). Between grades 1 and 3 there is an average increase in estimated vocabulary of about 9,000 words, or about 12 new words per day. This amounts to about 1 new word every 90 waking minutes! Many of these later-acquired words are derived words, such as "sadness" from "sad," the meaning of which is largely determined by the root word ("sad"). Nevertheless, language learning is a very significant task, confined to the childhood years. It therefore is not entirely surprising that the functional organization of the brain for language differs between the child and the adult, and that large, bilateral cortical regions are involved early on.

Restriction of language processing as well as processing of other tasks to smaller cortical areas in the adult may provide a functional advantage through reduction in unnecessary duplication, leading to improvement of the signal to noise ratio, to increase in the speed of processing, and to greater efficiency (decreased energy expenditure for performance of the same task). In the case of language learning in the child, the brain regions in the right cerebral hemisphere that are no longer needed for the processing of simple language functions may become specified for other, late-developing functions, such as music learning (Zatorre et al., 1994).

Left-sided lateralization of language for the simple word recognition task studied by Mills and colleagues (1993, 1997) occurred quite early, at about the age of 20 months, long before the end of the period of plasticity in language functions that is thought to depend on the ability to organize these functions in the right hemisphere, which has been estimated to occur between about the age of 8 years and puberty. What is the explanation of this discrepancy? It is likely that localization of language on the left side is a gradual process and that more difficult language tasks lateralize to the left hemisphere at later ages. Evidence from imaging studies suggests that some right cortical activation during processing of language tasks persists in normal adults.

The clinical and evoked potential studies provide confirmation that cortical representation at first tends to be diffuse, with overlapping functions rather than with strict cortical localization. As development progresses, there is increasing restriction of cortical representation, leading to specification of cortical regions for single functions. In other words, during development the cerebral cortex appears to shift from a distributed system to a more modular one. Development may bring more efficient processing by functional specification of cortical areas, but this may be at the expense of elimination of alternate routes of processing. Such alternate routes (for example, right hemisphere pathways for language processing) may be an important substrate for plasticity of the immature brain. Increased efficiency of processing in the adult may be bought at the expense of decreased plasticity.

There may be other explanations for the discrepancy between the age at which evoked responses to simple language tasks lateralize to the left and the age at which language functions can no longer be organized on the right side after left-sided brain damage. An attractive hypothesis is the following: Neural circuits or networks that are no longer utilized may stay in place for a time. Synaptic strength (the amount of depolarization of the post-synaptic membrane produced by an afferent volley) may decrease gradually. During this period, the unused circuits may be reactivated by the facilitation of synaptic transmission that is produced by stimulation of the system (Chapter 2). Only after a network has been unused for a long period (measured in years) may there be actual disappearance of synaptic connections. Such a sequence would explain many of the examples of neural plasticity that have been discussed in previous chapters: For example, non-alternating strabismus in infancy causes amblyopia or blindness of the squinting eye. This is associated with a decrease in synaptic input to the portion of the visual cortex that receives input from the squinting eye. However, the visual deficit can be reversed for several years by forcing the child to use the squinting eye, as by patching of the dominant eye. This reversal of amblyopia suggests that at least some connections from the squinting eye to the visual cortex were still present, and that these could be reactivated by visual stimuli presented preferentially to the amblyopic eye. An alternative hypothesis, which may be equally likely, involves the formation of an entirely new system of connections between geniculo-cortical axons and neurons in layer 4 of the visual cortex, perhaps by utilization of previously unspecified synapses.

PET and fMRI studies of language localization in normal adults and in subjects with early left-sided brain damage provide additional data relative to the cortical localization and plasticity of language functions. The results have varied somewhat, depending on the imaging technique and on the tasks used. The fMRI and PET study of responses to a simple semantic task (listening to single words) in right-handed normal subjects has shown primarily left cortical activations (Petersen et al., 1988; Binder et al., 1997).

These have included classic receptive language areas (such as Wernicke's area), but in addition there was activation of other left temporal and parietal areas, including the angular gyrus, and of large left prefrontal regions, including but extending beyond Broca's area. Right cerebellar activation also was noted.

Depending on the language task, right cortical activation may also be seen in normal adult subjects. For example, a PET study using oxygen$^{15}$ $H_2O$ as the radioactive isotope showed bilateral temporal activation in a task consisting of listening to a sentence. A sentence generation task activated large left frontal areas in addition (Muller et al., 1997). The imaging studies in normal adult subjects therefore show more widespread activation of the cortex during language tasks than would have been predicted by the classic model of speech localization. Moreover, there appears to be much individual variability (Ojemann, 1991). Pujol and colleagues (1999) studied cortical activation on fMRI scans taken when university students were performing a language task. A word generation task was used, in which the subjects were asked to "silently articulate" as many words as they could that started with a given letter. This task produced frontal lobe activation in the cortex that surrounds the inferior frontal sulcus, including Broca's area, the dorsolateral prefrontal cortex (Brodmann areas 46 and 9), and the premotor cortex. There was marked variation in laterality of activation: this was especially true in left-handed subjects, 58 percent of whom had strong left lateralization, with the remainder showing various degrees of bilateral activation. Only 2 percent of the left-handers showed strong right lateralization. Even in right-handed subjects, 18 percent showed some degree of bilateral activation.

It therefore is not surprising that imaging studies of language localization in subjects with left brain lesions also show marked individual variability (Muller et al., 1999). However, a trend is noted toward greater reorganization of language in the right hemisphere in subjects with a history of stroke prior to the age of 5 years than in those who had a stroke after the age of 20 years, as predicted from the clinical data (Muller et al., 1997b). By and large, the reorganization of language in the right hemisphere appears to occur in cortical regions that are analogous to the language areas on the left side.

The contribution of fMRI to the study of language localization in infancy and childhood has been limited because such studies are difficult to perform in young children. The youngest ages at which language localization has been assessed by fMRI have been 7–9 years. In this age group, language processing appears to take place in the left cerebral hemisphere in a way similar to the adult pattern (Benson et al., 1996).

There may also be gender-related differences, with greater contribution of the nondominant hemisphere to language processing in women (B. A. Shaywitz et al., 1995). Data from fMRI studies by Shaywitz and his coworkers (B. A. Shaywitz et al., 1995; Pugh et al., 1996) provide interesting information concerning gender-related differences in the laterality of cortical representation of some aspects of language. Subjects were asked to sound out two strings of nonsense words, and to determine whether the two strings rhymed. This phonological task produced differences in the degree of lateralization of function between female and male young adult subjects. Females tended to have bilateral activation of the inferior frontal cortex (Brodmann areas 44 and 45), while activation tended to be restricted to the dominant side in males. A host of questions arises: Is bilateral activation in females associated with superior phonological ability? Does it correlate with lower performance in other functions, such as mathematical ability and spatial orientation, related to crowding effects? Are these differences hormonally regulated? Careful correlation of data from fMRI and from neuropsychologic and endocrine studies could provide answers to these intriguing questions.

The studies reported by B. A. Shaywitz et al. (1995) are of importance in that they suggest effects of the internal environment on the organization of higher cortical functions. These effects, including hormonal effects on brain development, have been relatively neglected. The hypothesis to be tested is that estrogens mediate the maintenance of bilateral cortical representation of language processing. This question will be taken up again in the discussion of reading in Chapter 7.

In the processing of the more challenging language tasks, and even more so in the representation of higher cognitive tasks, one gets the impression that cortical processing becomes less clearly localized than it is in simpler functions, such as sensory processing and motor control (Just et al., 1996; Derbyshire et al., 1998). The imaging findings are consistent with processing of higher cortical functions as a distributed network that involves parallel connections between multiple cortical regions (Mesulam, 1990; Bates, 1993; Small, 1994; Elman et al., 1996; Y. H. Kim et al., 1999).

## The Superiority of the Left Cerebral Hemisphere for Language

One may ask why, if both hemispheres are capable of processing language functions in the young child, is it the left hemisphere that becomes dominant for language in most humans. This fact suggests that the left hemisphere may have an inherent superiority over the right when it comes to language functions. There is some support for this from neuroanatomy: the planum temporale, specifically the site of Wernicke's area, is larger on the left than on the right (Geschwind and Levitsky, 1968; Wada et al., 1975; Galaburda, 1984). This difference is evident already in the neonatal brain (Witelson and Pallie, 1973; Chi et al., 1977; Best, 1988), which suggests that it is genetically determined. There may be competition for language representation between the two hemispheres. In such a competition, the left would have the advantage of a larger region available for language processing. Size of a cortical region appears to be a factor in its functional capacity. This could be due either to the increased size of the neurons, due to more complex dendritic and synaptic development, or to an increase in the number of the vertically arranged columns or functional modules that are the building blocks of the cerebral cortex. These issues will be taken up again in the discussion of Einstein's brain in Chapter 7.

## Environmental Effects on Language Development

A view that has been widely held sees language development to a large extent as an innate human function that has only limited dependence on environmental input (Pinker, 1994). This is thought to be true especially for grammar. In support of this hypothesis is the finding that grammar tends to show little variation from one language to the other. It has also been claimed that grammatical competence emerges early and very rapidly, too fast to be learned by experience (Crain, 1991). The lively controversy surrounding this claim is illustrated by the numerous responses published at the end of Steven Crain's 1991 paper.

Recently, it has been found that language development, including the development of grammar, is significantly influenced by environmental input. During the second and third years of life, the amount that the mother talks to the child is highly correlated with the child's vocabulary size. The most likely explanation is that the amount of speech a child experiences influences language development in a positive way (J. Huttenlocher et al., 1991). In this type of study, it is very difficult to separate genetic from environmental factors. However, a genetic explanation of the findings appeared unlikely since the vocabulary and grammar of the mothers who talked more to their children were not significantly different from those of the less talkative mothers. In addition, the amount a mother talks is not a measure of her language ability. However, there might be other genetic differences, as yet unidentified, which result in more talking by the mother—and also in a larger vocabulary in her child—independent of the amount of input.

This possible confound is avoided in an ingenious experimental design in which school effects on language development are assessed (J. Huttenlocher et al. 1998). School effects become more important than mother's speech in all aspects of language learning once the child reaches the age of 5 or 6 years. In this study, growth in language functions, including grammar, was compared during the months of school and the months of summer vacation. If language growth proceeded along a genetically determined trajectory, independently of environmental input, the developmental curve should not be influenced by school experience. Instead, substantially more growth of all aspects of language was noted during the school year than during the summer vacation period, during which there is relatively little growth (Figure 6.2). This is true even for the aspects of language that had been widely held to be genetically determined (Crain, 1991), and in particular for grammar. The growth curves reproduced in Figure 6.2 provide a beautiful example of the effects of the environment on the cognitive growth of the child and—implicitly—on the effect of input on the functioning of the developing brain.

A recent study by Hoff-Ginsberg (1998) lends support for the presence of important environmental effects on language acquisition. Hoff-Ginsberg compared language development in first-born children to that in later-borns. The first-born children were more advanced in lexical and grammatical development, a difference that appeared to correlate with an increased amount of the mother's speech to the first child. On the other hand, conversational skills were more developed in children with higher birth order, a finding perhaps related to greater practice of conversation in a larger house-

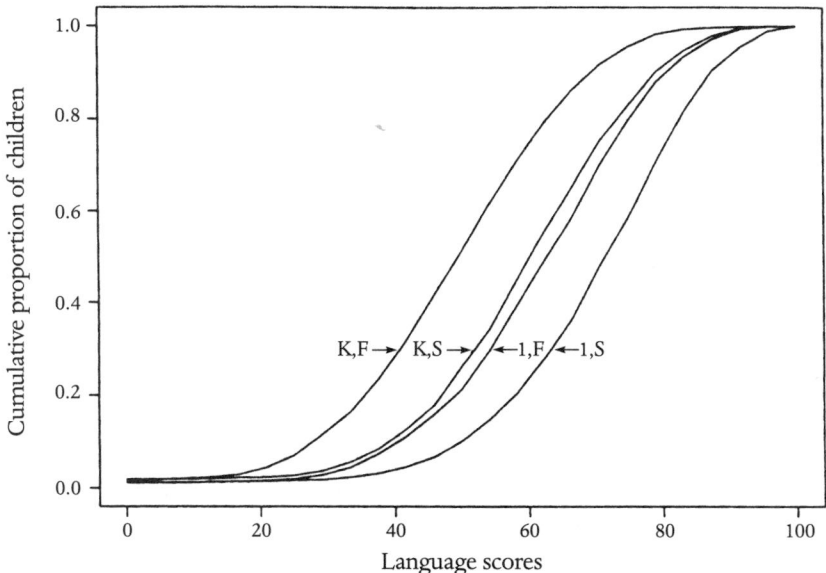

Figure 6.2. Growth in language functions in relation to school attendance and summer vacation. K,F = kindergarden year, fall; K,S = kindergarden year, spring; 1,F = first grade, fall; 1,S = first grade, spring. (From J. R. Huttenlocher, Levine, and Vevea, 1998, fig. 2A, p. 1022; reprinted with the permission of the Society for Research in Child Development.)

hold, with multiple opportunities to interact. There also was an effect of socioeconomic status (SES) on language development. These findings are difficult to reconcile with the view that development of language, in particular of grammar, occurs according to an innate program, independently of environmental exposure. This is not to deny the existence of genetically determined factors that influence language development. Genetic factors have been described especially in relation to abnormal language development (Hurst et al., 1990).

As yet, we know little about the changes in the organization of the cerebral cortex that underlie these developmental events. Evoked potential data, published by Helen Neville and her coworkers, and cited earlier in this chapter, indicate that children with advanced language skills have earlier restriction of word processing to the dominant hemisphere than do those with slow language development. This suggests that the advanced children have earlier development of more efficient language processing. At this point, we do not know whether this difference is due to environmental or genetic factors. However, this question could now be approached by the noninvasive study of groups of young children with different language abilities and/or different environmental input, using evoked potential or fMRI techniques.

### Cerebral Localization of Language in the Congenitally Deaf

Neville and coworkers (1998) used fMRI to study the localization of American Sign Language (ASL) and of English in congenitally deaf subjects. All were native signers of ASL, which they had learned starting in infancy, as a first language. English was acquired late through the visual modality, that is to say by learning to read. The test stimulus consisted of silent reading of a paragraph, presented either as an English text or in the form of ASL communicated by a native signer. ASL had a cortical representation in the language areas of the left cerebral hemisphere, with additional activation in homologous areas of the right cerebral hemisphere, which was more extensive than the right hemisphere activation by English that is seen in normal, hearing subjects. It therefore appeared that ASL was processed in a way similar to how other languages are processed, with additional activation of regions in the nondominant hemisphere, probably reflecting the visual-spatial aspects of ASL analysis.

More surprising, brain areas activated during English reading did not include the language areas in the left cerebral hemisphere. The most consistently activated areas were in the right middle and posterior temporo-parietal cortex, but there appeared to be considerable individual variability. In contrast, hearing subjects who—because of the deafness of their parents—had learned both English and ASL as infants showed normal language localization for English in the left hemisphere. In these subjects, English and ASL appeared to share the language areas for cortical processing. ASL in addition activated right hemisphere regions, as in the congenitally deaf. The findings suggest that by late childhood, when congenitally deaf persons are likely to learn English by learning to read, the language areas of the brain have lost the plasticity needed to accommodate a second language. Instead, the second language becomes represented in other cortical regions. This also appears to be true for the learning of second languages in individuals with normal hearing.

Second language acquisition provides an interesting example of the superiority of the immature brain for certain types of learning. Young children exposed to a second language (L2) seem to learn L2 with little effort (Johnson and Newport, 1989; Flege et al., 1995; Weber-Fox and Neville, 1996). While there are some claims that acquisition of a second language during childhood requires as much or more time and effort than second language learning in the adult, there is little question that language learning in the child is more perfect than it is in most older people. Second languages are learned flawlessly by young children, who speak them without accent and with normal grammar. The "window of opportunity" for accent-free language learning extends to early adolescence, with some individual variation. Lenneberg (1967) gives a time window of 2 years of age to puberty for optimum language learning. The time window for second language learning appears to depend on the degree of difference between the two languages. For similar languages, such as English and German, the upper limit appears to be in early adolescence (Newport, 1990). For dissimilar languages, such as English and Korean, L2 learning may already be imperfect when started at the age of 3 years (Weber-Fox and Neville, 1996).

The upper limit of the time window for accent-free second language learning coincides approximately with the age at which cerebral metabolic rate and synaptic density in the language areas of the brain decrease from their high, early childhood levels to the adult range, during late childhood to early adolescence. These events may be causally related, but there is no direct proof of this at present.

Recent data from functional imaging studies suggest that L2 is represented differently in the cerebral cortex, depending on age at the time of acquisition. Kim and colleagues (1997) used fMRI to study the cortical representation of L1 and L2 in relation to the age at acquisition of L2. All subjects were highly proficient in L2. The task used involved "silent speech," to describe occurrences during the previous day. Subjects who learned L2 in infancy were compared to those whose first exposure to L2 came after adolescence. In the subjects with exposure to L2 in infancy, the brain areas activated during silent speech in L1 and in L2 were identical, in Wernicke's and Broca's area in the left cerebral hemisphere. In subjects who learned L2 late, activation in Wernicke's area was the same in L1 and in L2. However, the region in Broca's area activated by L1 was not activated during L2. Instead, L2 was represented in a frontal cortical region adjacent to L1. It appeared that expressive language functions in L2 could no longer be integrated into the normal Broca's area in the late learners of L2, while such integration was still possible for receptive language. This is of interest, since expressive language (accent and use of grammar, functions presumed to depend on Broca's area and on the frontal cortex near Broca's area) tends to be impaired when L2 is learned late, while language comprehension, a function of Wernicke's area, may be flawless. The findings also suggest that areas of the brain not usually used for a specific task, such as speech, may become enlisted for this task under special circumstances. Examples of this in the sensory and motor cortex have been discussed already in previous chapters. The data from second language learning suggest that this may be less perfect when regions that are normally not specified for language become enlisted. This may represent an effect of crowding, related to the processing of multiple functions in one cortical area.

More recent work on the cortical representation of L2 has resulted in somewhat different but not necessarily contradictory results. Perani and colleagues (1998) compared early and late learners of L2, with both proficient and imperfect language learners in both groups. Cortical activity during a language task appeared to be related to proficiency rather than to the age at which the language was learned. Subjects with high proficiency in L2 had equal representation of L1 and L2 in Broca's area. Subjects with low proficiency had L2 representation outside Broca's area. However, the question whether the imperfect speakers of L2 had reached their best possible performance level was not addressed. In other words, representation of L2 outside Broca's area may represent a stage in the learning of L2 in the young child, and L2 may become incorporated into Broca's area in the same child as language learning progresses. Nor do we know the total incidence of proficient versus imperfect speakers of L2 as a function of age in the population from which the subjects were drawn. Organization of L2 outside Broca's area may be associated with imperfect mastery of L2, independently of age, but perfect mastery of L2 may be uncommon in the late learners, while it occurs frequently in early learners.

The loss of the ability to learn to speak a language without accent may be related to maturational changes in the ability to make certain sound discriminations. Young infants have equal ability to discriminate similar sounds, such as "r" and "l." However, the adult is unable to make certain sound discriminations that are not used in the native language (Tees and Werker, 1984; Werker and Tees, 1984). Between the ages of 6 months and 1 year, sound discriminations that are of importance in the language the infant is exposed to become sharpened, while sensitivity to nonnative speech contrasts diminishes (Kuhl, Williams, et al. 1992; Kuhl, 1994; Kuhl, Andruski, et al. 1997). This loss in the ability to discriminate certain speech sounds may underlie the inability of a person to learn a second language without accent after early adolescence. Nonnative sound discriminations important for L2 would have to be practiced from an early age for an adult to be able to learn L2 without accent. The motto "use it or lose it" is sometimes applied to this type of developmental change. Evidence from second language learning suggests that the ability to make nonnative speech sound discriminations can be recovered until late childhood. Beyond that age second language acquisition tends to become imperfect, with persistence of an accent that depends on the early language experience of the learner. The upper age limit for retrieval of speech sound discrimination by practice is considerably higher than is the age at which unused sound discriminations usually disappear. This difference may represent another instance of long persistence of unused cortical circuits, and of their reactivation by practice.

Developmental loss of abilities occurs in a number of different areas of cognitive function. Some examples have been discussed previously in this chapter, where possible mechanisms for the loss of functions during development were presented. In the area of second language learning and retention, developmental loss is a major factor. Young children rapidly lose a second language if it is not practiced. Bilingual families are well aware of the fact that they will need to speak the second language at home for the child to remain proficient.

### Bird Song

Bird song, while clearly different from human language, nevertheless provides some interesting parallels as far as plasticity is concerned (Doupe, 1996; Doupe and Kuhl, 1999). The basic ability to produce a song is present in every male bird. However, the nature and intricacy of the actual song are environmentally determined. The strategies that are used differ in different species. Finches have been studied most thoroughly. Young birds learn their songs from their father or some other tutor, even from tape-recorded bird songs, during an "impressionable period." Feedback from the bird's own songs as well as the experience of a tutor's songs appears to be necessary for the shaping of the young bird's own songs (Solis and Doupe, 1997, 1999). The impressionable period varies in different species, both in the age of the bird (starting as young as age 7 days from hatching) and in the duration of the impressionable period (3 to over 100 days)(Konishi, 1995). The young bird does not actually produce the memorized song during the impressionable period. However, a bird that has heard the song during this period will reproduce it faithfully after it reaches sexual maturity. Birds prefer learning the songs of their own species, but will learn other patterns if deprived of species-specific song patterns during the impressionable period. Birds deprived of all semblance of bird song during the impressionable period will produce very abnormal, truncated vocalizations, sometimes referred to as "innate bird song."

Environmental input during a critical period therefore is as important a prerequisite of bird song as it is of human language. The actual production of songs is regulated by the male sex hormone, testosterone. Castrated birds will not sing, but will do so when injected with testosterone. This finding also may have relevance to human language acquisition. Hormonal effects may regulate language development and cerebral localization of language in humans (B. A. Shaywitz et al., 1995). It is probably no accident that major changes in language organization and plasticity occur at puberty. However, our present knowledge about hormonal effects on bird song much surpasses our information about hormonal effects on human language development.

In most species of birds, songs remain relatively stable throughout the life span. However, canaries are able to modify their song annually, with the appearance of a new season. The annual renewal of songs in canaries is associated with the formation of new neurons in areas of the brain important for bird song learning (Nottebohm, 1985). The formation of new neurons has also been found in the hippocampus of chickadees, a brain area important for the memory of song patterns (Barnea and Nottebohm, 1994). This formation of new neurons also occurs seasonally and is greater in birds living in the wild than in captive birds. The hippocampus is one of an increasing number of regions in which the growth of new neurons has been demonstrated after birth in other species, even in primates (Gould et al., 1998,

1999b). Very recently, the formation and cortical migration of new neurons in adult primates has been shown to occur in areas related to cognitive, including language, functions (the prefrontal, posterior parietal, and inferior temporal cortex), but not in the primary sensory cortex (Gould, Reeves, Graziano, and Gross, 1999).

It appears that a number of parallels exist between bird song and human language, which could be exploited for the study of critical periods, hormonal effects, and the possible functional significance of adult neurogenesis.

## *Summary*

The plasticity of the immature brain relative to language functions is remarkable and in sharp contrast to what is seen in the adult brain. This is clearly evident in the case of lesion-induced plasticity. Lesions that destroy the classic speech areas in the perinatal period are compatible with development of nearly normal speech, in contrast to similar adult lesions, which produce permanent aphasia. The ability to develop language after perinatal dominant hemisphere lesions is thought to be secondary to the organization of language functions in the nondominant (right) hemisphere. Recent evidence suggests that this is related to the initial normal bilateral representation of language.

Plasticity in language development has also been demonstrated in the normal child. The importance of input, such as the mother's speech and school effects, has recently been established. In this work, it has been important to distinguish environmental effects from unequal endowment due to genetic factors. This has been accomplished by comparison of the developmental trajectory during the school year and during summer vacation. Input effects have been shown both for vocabulary and for the development of grammar. The plasticity of language functions in the immature brain also is demonstrated in relation to second language learning. This appears to be more effortless if the second language is introduced when the learner is young, and it is more perfect as far as accent and grammar are concerned. Recently, it has been shown that high proficiency in the second language is associated with its representation in the primary speech areas, where it shares the area of representation of the native language. Late and imperfect acquisition of the second language are associated with the formation of a new speech area adjacent to Broca's area, which suggests that during maturation Broca's area loses the ability to process more than one language.

Bird song provides interesting parallels to human language, in that there is a clear environmental influence on its form and complexity. Furthermore, there are clearly defined critical periods, but these, as in humans, end gradually rather than abruptly. Hormonal factors are important in the development of bird song, as they may also be in some aspects of human language, especially reading (Chapter 7). Songbirds therefore provide a model in which some aspects of the plasticity of human language development can be critically evaluated, and genetic, hormonal, and environmental factors can be assessed.

# 7

## PLASTICITY IN ELECTIVE BRAIN FUNCTIONS

> The frontal cortex comprises a third of the human brain; it is the structure that enables us to engage in higher cognitive functions such as planning and problem solving. What are the processes that serve as the building blocks of these higher cognitive functions, and how are these implemented in frontal cortex?
>
> Edward Smith and John Jonides, 1999

Little is known about cortical representation and plasticity of the so-called higher functions, including abstract reasoning, such as is needed for the understanding of higher mathematics and physics, and the executive functions such as judgment, values, motivation, and planning (Sternberg and Powell, 1983). Even less is known of the anatomical substrate of creativity, the creation of new knowledge and of works of art. These as a group represent "elective brain functions." They differ from the basic or obligatory cortical functions such as vision, hearing, and locomotion in that they depend completely or nearly so on specialized teaching and learning. What is perceived as correct behavior with respect to values, judgment, and motivation depends to a large extent on the culture in which a person grows up and lives. Some higher cortical functions also require creative drive. This appears to be to a large degree self-generated. In other words, we at some point in life become able to influence the function of our own brains by self-generated input to the brain. An example is the intense practice of a concert violinist, who has developed the drive to perfection in performance. The violinist is able, through his or her own actions, to change both the anatomy and function of his or her own cerebral cortex (see below).

Speech, as we have seen in Chapter 5, has a somewhat intermediate position between obligatory and elective cortical functions: the acquisition of a specific language depends on exposure, that is, on the language the infant hears, and the complexity of language that is achieved is dependent on exposure and training. However, a basic spoken language develops in every child with normal hearing and intelligence who has some exposure to a language. In contrast, another major language function, reading, requires special teaching to develop at all. A discussion of reading therefore is included in the present chapter on higher cortical or elective functions.

Higher cortical functions are represented primarily in two cortical regions: the prefrontal cortex and the posterior parietal cortex. These are cortical regions that develop slowly and that have the highly complex dendritic development that appears to be necessary for the processing of complex information. Most of these functions are distinctly human, and information concerning them is dependent on observation and study of humans. FMRI is emerging as an especially important technique for unraveling the neural basis and malleability of these functions. The major dependence of higher cortical functions on environmental input suggests that plasticity is high in these functions, but we have little specific knowledge concerning this except in reading and in the development of musical ability.

### Reading

Reading, like speech, is a uniquely human function. It differs from receptive oral language and speech in that extensive training is required for its acquisition, even for its basic aspects. All normal children exposed to a spoken language will acquire the language, albeit with varying degrees of proficiency. One can therefore take a point of view that much of language is innate in the human brain, and especially that the structure of language (syntax) is inborn. In Chapter 6, evidence was presented that this is not entirely so, and that all aspects of development of spoken language are subject to modification by the environment. It is doubtful that an oral language would emerge without some exposure to a human language. The limited available data on so-called feral children suggest that it would not (Curtiss, 1977). However, the desire to communicate is present in every normal infant and—given the availability of another human that is able to communicate—spoken language will develop.

As far as written language is concerned, there can be no question that environmental input is of paramount importance. In the absence of considerable training and exposure to written materials the child will not learn to read and write, but will grow into an illiterate adult. There are many examples of societies in which the entire population or the majority are illiterate. Reading in this respect differs from spoken language, which is present in all human population groups. It may be considered an elective rather than a universal human function.

There appear to be important differences between the cerebral organization of oral and written language. The child acquires oral language in the second year of life, at a time when language areas in the brain mature, and when synaptic density in these cortical areas approaches the maximum. Reading, on the other hand, is not acquired until the age of 6–7 years in most children. This is several years after the occurrence of synaptogenesis, which correlates well with the emergence of basic cortical functions such as the processing of sensory stimuli, voluntary motor activities, and oral language, but not with the development of elective functions such as reading. The data show no evidence that the left angular gyrus, or the other brain areas that are thought to be important for reading, develop later than other language areas. What, if anything, is the function of these cortical areas prior to school age? We have no knowledge concerning this, but would now be able to design studies that could test different hypotheses. For example, an attractive hypothesis is that the angular gyrus is required for other language functions prior to the age of 5 years. The likelihood of this is increased by the fact that the angular gyrus abuts Wernicke's sensory speech area. It would be consistent with other evidence that suggests more diffuse functional representation of spoken language in the immature brain, with gradual restriction of areas of representation during childhood. Restriction of functional localization of early-developing, primary functions may free up cortical space for later development of elective functions such as reading, music, mathematics learning, map reading, and so on.

Interesting gender-related differences exist in reading ability, with girls, on average, acquiring reading skills earlier than boys. Dyslexia, while apparently equally common in boys and girls (S. E. Shaywitz, B. A. Shaywitz et al., 1990), tends to be less severe in girls. The question arises whether these differences are associated with gender-related differences in cortical representation of reading. Pugh and colleagues (1996) compared cortical activation on fMRI scans during a reading task in men and women. They compared the cortical areas important for analysis of the component processes in reading (visual or line judgment, orthographic or letter recognition, phonological or word sounds, and semantic or word meaning), using a multiple subtraction technique. The results in female and male subjects were analyzed separately. Activation patterns were widespread in all subjects, but there was increased bilateral activation in females. An important and as yet unresolved question is whether the cortical representation of reading in boys is at first bilateral and—if so—at what age predominant left hemisphere localization normally appears.

The question arises whether the sex differences in cortical representation of reading are hormonally mediated. Cortical representation of some other higher cortical functions appears to be affected by estrogen levels.

The question of estrogen effects on higher cortical functions has recently been pursued in an fMRI study by S. E. Shaywitz, B. A. Shaywitz, and their colleagues (1999). Two groups of postmenopausal women, one on estrogen replacement therapy and one off it, were tested in a double blind, crossover design. Areas of brain activation during verbal and nonverbal memory tasks, as detected by fMRI, were determined for both groups. Comparison of the two groups showed increased right cortical activation in subjects on estrogen therapy. This difference may have functional significance, since several studies have shown an improvement in verbal memory in postmenopausal women treated with estrogen (Sherwin, 1997). It therefore appears that significant hormonal effects on cognitive functions and on their cortical organization may exist in humans. This is a subject of considerable investigation and speculation at the present time. As indicated in Chapter 6, hormonal effects are very important in the initiation of bird song, and this finding provides an animal model in which hormonal effects on language-related functions can be studied.

### Musical Ability

The ability to appreciate and compose music and to play a musical instrument represents a set of functions in which early input appears to be especially important. Most famous musicians, including composers and performers, have had training that started early in childhood. Many were proficient instrumentalists and composers already in childhood and adolescence. For example, Mozart learned to play the harpsichord at the age of 3–4 and began to compose music at the age of 5. Rossini produced remarkable compositions during childhood. Mendelssohn was an accomplished pianist by the age of 9, when he impressed Goethe with his playing, and Schubert wrote several of his major compositions as a teenager. In these cases, it is difficult to separate genetic from environmental factors. In other words, there may be "music genes" that to a large extent determine whether a given individual has the capacity for superior musical composition and/or performance. The Bach family, in which there were multiple family members in several generations with superior musical ability, has often been cited as an example of genetic influences in musicality. However, the same pattern of superior musical achievement in multiple family members could also have been the result of extensive early exposure. In almost all histories of famous musicians there is an adult (the father in Mozart's case) who took an intense interest in the musical training of the child. The success of the Suzuki method, in which playing of the violin is started in the preschool years, also suggests the importance of early training. Yet—as many a parent has noted—early exposure to music does not guarantee the development of high proficiency. There appears also to be an important factor of natural endowment.

One aspect of learning to play the violin is the development of remarkable skill in movement of the fingers of the left (nondominant) hand, in response to signals received from other cortical regions, probably in the nondominant parietal cortex. These are pathways that do not develop without extensive early training, probably before the onset of adolescence. Proficiency as a violinist has been reported to be associated with a change in the area in the right motor cortex that controls the finger movements of the left hand, the hand that depresses the strings on the fingerboard (Elbert et al., 1995). The technique used in this study was that of magnetoencephalography (MEG). The size of the magnetic response to left finger movements was found to be larger in violinists than in nonplayers. The responses were measured in adults, and were found to be largest in those subjects that had started violin playing early, prior to adolescence. There may, therefore, be a time window for this effect, but one that closes gradually. The effect of violin practice on the size of the motor hand area provides one of the pieces of evidence that suggest a correlation between size of a cortical area specified for a given function and competence in that function.

*Table 7.1.* The age at which musical training is started and the acquisition of absolute or relative pitch. A.P. = absolute pitch; R.P. = relative pitch. The numbers in parentheses are the Standard Error of the Mean (SEM). (Based on data from Zatorre et al., 1998.)

| Group | Age (years) | Musical experience (years) | Age at start of music training |
| --- | --- | --- | --- |
| A.P. | 24.5 (1.3) | 18.2 (2.0) | 6.3 |
| R.P. | 24.8 (2.1) | 13.1 (2.2) | 11.7 |

Jancke and colleagues (1997) found that early learning of piano playing increases finger tapping performance, especially in the left hand, which leads to a decrease in left-right differences in this task. This effect was seen when music training was started early in childhood, but did not correlate with the duration of the musical training, which again suggests a time window.

Another time window related to musical competence is suggested by data presented by Zatorre and colleagues (1998). This study investigated the presence of absolute pitch, the ability to identify the pitch of any sound without reference to another sound and to produce a given musical note on demand. Absolute pitch is not present in nonmusicians and is seen in a minority of musicians. It develops primarily in musicians whose musical training began early. In the report by Zatorre and colleagues (1998) the average onset of musical training in musicians with absolute pitch was at the age of about 6 years, while the average onset of musical training in musicians without this skill was at the age of about 12 years (Table 7.1). The data suggest that acquisition of absolute pitch is unlikely unless musical training is started before the age of 12 years. Hirata and colleagues (1999), Pantev and colleagues (1998), and Pantev and Lutkenhoner (2000) used MEG to study neural activity in the auditory cortex in musicians with absolute pitch. Nonmusicians served as controls. Musicians with absolute pitch were found to have activation in the posterior auditory cortex by tones that was not present in nonmusicians. These studies do not clarify the issue of whether the difference in activation pattern was secondary to training or whether it reflects an inherent difference related to musical talent.

The possible effects of early musical experience on cerebral organization have been a subject of recent study. Already in the fetus, a soothing effect of playing Mozart has been reported, but the long-term effects of such exposure are not well substantiated. There have also been reports that listening to music has short-term effects (for a matter of hours) on the performance of some higher brain functions in adults. It has been reported that listening to Mozart, but not to repetitive musical sounds, improves performance on spatiotemporal tasks that are important for mathematics and chess playing—the "Mozart effect" (Rauscher, Shaw, and Ky, 1993, 1995; Rauscher et al., 1997). In this regard, it is of interest that spatial tasks and music are both represented in the same general cortical region, the nondominant parietal cortex posterior to the postcentral gyrus. The question arises whether processing of one task by a cortical area with multiple functions may prime this area for performance of other tasks that are also represented in the same area. Such a mechanism, if it exists, would be expected to be especially important in the executive functions represented in the prefrontal cortex, a cortical region that appears to be capable of processing a large variety of cognitive functions, including reasoning, judgment, and planning, without clear localization. However, the very existence of the "Mozart effect" is somewhat in doubt, since attempts to replicate have not always been successful, and since the interpretation of the findings has also been challenged. For example, Nantais and Schellenberg (1999) confirmed the "Mozart effect" when the effect of playing Mozart was compared with that of sitting in silence. Not surprisingly, the effect was not specific to the playing of Mozart: a piece by Schubert had the same effect. More important, the "Mozart effect" disappeared when the playing of Mozart was compared to listening to a short story. Subsequent performance on a spatial-temporal task (the paper folding and cutting task on the Stanford-Binet Intelligence Scale) appeared to be related to the subject's preference (listening to a story versus listening to music), rather than to exposure to Mozart. The findings suggest either of two explanations: that sitting in silence impairs subsequent performance on spatial-temporal tasks, or that prior exposure to pleasant entertainment improves such performance. The effect may well be related to the influences of mood and arousal on test performance.

### The Executive Functions

Functional localization appears to become less precise as cortical functions become more complex, as in mathematical and logical reasoning and in the so-called executive functions. These may be represented in multiple brain regions, and there may be individual differences in cortical processing. The existence of multiple circuits that are involved in mathematical thinking is supported by evidence derived from fMRI studies. Dehaene and colleagues (1999) found activation in large cortical regions, including classic language areas and the bilateral parietal cortex, during the processing of mathematical tasks. Different tasks appeared to have different patterns of activation.

Executive functions are thought to be to a large extent represented in the prefrontal cortex. This knowledge is based on the study of patients with bilateral frontal lesions, such as prefrontal lobotomy or leukotomy (Stuss and Benson, 1986), and more recently on the results of functional imaging—fMRI and PET—during performance of reasoning tasks (Smith and Jonides, 1999). However, the exact localization of executive functions has been difficult to pinpoint. Unilateral prefrontal lesions often do not produce any obvious deficit, whether the lesion is on the right or on the left. Neurosurgeons refer to these regions as "silent areas": regions whose involvement by tumors or other focal lesions or focal resection fails to produce a deficit. Deficits appear only when both frontal lobes are damaged. It appears that both the right and left prefrontal cortices are organized to subserve executive functions and that the two sides may be nearly "equipotential" throughout the life span. However, some predominance of activation in the left frontal lobe during verbal reasoning tasks, and in the right frontal lobe during spatial reasoning tasks, has been found by PET (Smith and Jonides, 1999). PET studies have also been carried out during tasks requiring attention to a given set of information and simultaneous inhibition of irrelevant information, thought to be mediated in the prefrontal cortex. Such tasks are provided by the Stroop test, in which a subject needs to learn new, wrong names for colors (such as "red" for "blue"), and to inhibit attention to the correct color name. PET scanning during performance of this test has shown consistent activation of the anterior cingulate gyrus, suggesting that this may be an important cortical region for the resolution of "cognitive conflict" (Derbyshire, Vogt, and Jones, 1998). However, there were in addition large areas of activation that were different from one subject to the other, which suggests that different cortical regions may be enlisted for the processing of the same task.

In the processing of executive and reasoning tasks the prefrontal cortex may function as a distributed system, with parallel processing of information, and without specific functional modules. This is an active area of present research that promises to yield much new information concerning the development and organization of higher cortical functions in humans.

What appears to emerge is that for prefrontal lobe functions, the search for a simple map of representation of specific functions in specific cortical regions may be futile. A more likely model appears to be one of involvement of the same cortical areas for multiple functions. The pattern of activation in relation to a given task may differ in different persons. There may be more than one strategy for execution of the task even in the same person. Some evidence of normally occurring variability in cortical processing has been found even in basic tasks such as speech and reading. Gender differences in brain activation during reading have already been discussed. Some adults with normal speech have bilateral activation of the cerebral cortex during language tasks, such as word generation (Pujol et al., 1999), while in the majority there is predominant left-sided activation. These examples provide evidence that different strategies for neural processing may be employed to achieve similar functional outcomes.

## Environmental Effects on Intelligence

Intelligence as measured by IQ tests has often been considered to be a fixed, genetically determined quantity that persists unchanged throughout the life span of a person. As has already been pointed out by Lashley (1951), this is in part determined by the nature of IQ tests, which are constructed so as to exclude items that are affected by environmental factors, such as school learning.

Nevertheless, there is evidence that the environment can affect IQ scores. This derives largely from studies of preschool enrichment programs. The Abecedarian program represents such a study. It shows that a high-quality, preschool intervention program, starting at birth and continued until kindergarden age, can affect childrens' IQ scores (Ramey and Campbell, 1984; Ramey and Ramey, 1998). By the age of 5 years, mean differences between children in the intervention program and randomly assigned controls were sizable, varying between 10 and 18 IQ points. These large effects were seen in children from socioeconomically deprived backgrounds (Ramey and Ramey, 1994). More recent evidence shows smaller differences in less deprived groups, suggesting a ceiling effect. After the age of 5 years, the control and early intervention program children attended the same schools and no further special support was provided. Long-term follow-up showed a gradual decrease in the IQ differences. By the age of 15 years, the early intervention and control groups still showed significant differences in IQ, but the mean difference had decreased to 4.6 IQ points. There also were significant differences in school achievement and in the need to repeat a grade (Campbell and Ramey, 1994, 1995). The results were similar to those obtained in the preschool (Head Start) programs of the 1960s (Darlington et al., 1980). These results have been variably interpreted as indicating the significant benefit of early intervention and the relative failure of such programs as far as long-term effects are concerned. The cup may be looked at as half full or half empty (Bruer, 1999). However, it appears quite amazing that significant differences between the groups persisted almost 10 years later, when one considers that all the children returned to a deprived environment, which exerted its negative effects, especially later in childhood and adolescence, when peer pressures become paramount. Both the increase in IQ seen by the age of 5 years and the later decrease in late childhood are examples of environmentally determined IQ effects. We have seen already in the discussion of second language learning, that a second language acquired early in life is quickly lost, unless practiced. The saying "use it or lose it" can be applied to intelligence as well as to second language retention. The synaptic plasticity present in the immature brain may make developmental loss of functions that are not practiced especially likely. The anatomical data suggest that adolescence may be a critical period during which many earlier-acquired functional circuits may be lost, unless they are utilized. In order to maintain the benefits of early enrichment programs, it may be especially important to provide intellectual stimulation, including review of previously learned material, during late childhood and early adolescence.

The effects of early intervention on IQ have been reported primarily in two groups of children. The first are children from deprived environments, exemplified by those in the Abecedarian program. The second are children at high risk of developmental handicaps for medical reasons. Infants born prematurely have been studied the most carefully. The most extensive study has been the Infant Health and Development Program (IHDP) sponsored by the National Institutes of Health. It shows that the long-term outcome for cognitive functioning in small premature babies is better when the children grow up in a nurturing environment (Brooks-Gunn et al., 1992, 1993). Premature infants raised in a deprived environment have a high incidence of significant cognitive, learning, and adjustment difficulties. Early childhood enrichment programs benefit this group of children. The IHDP provided early intervention from the age of 12 to 36 months. At the end of this period, children in the early intervention program had IQ scores up to 13 points above those in controls. The effect was seen primarily in infants from low socioeconomic and educational groups. There was limited benefit in children from middle-class environments, suggesting that such children are likely to receive the necessary stimulation in the home. As in the Abecedarian study, the IQ effect of early intervention gradually decreased during several years of follow-up. It appears most likely that maintenance of an enriched environment throughout childhood and adolescence would maintain these early gains.

What we can conclude from these studies is that long-term effects of early intervention are often subtle, and are seen mostly in children who live in a globally deprived environment. Single factors in the child's life, such as perceived substandard quality of day care, may have limited impact, as long as other factors such as home environment and parental interactions with the child are normal (Scarr, 1998).

### IQ Effects of Early Focal Brain Lesions

Psychological testing of children with unilateral perinatal brain damage shows a negative effect on overall intelligence (Perlstein and Hood, 1954). Comparison of subjects with left versus right hemisphere perinatal lesions, using standard tests such as the Wechsler Scale, has shown that both groups perform slightly better on language than on performance items (St. James-Roberts, 1981; Levine et al., 1987) (see Table 7.2). In the study by Levine and colleagues, the mean verbal IQ on the Wechsler Scale in children with perinatal left hemisphere lesions was 82, that in perinatal right hemisphere lesions was 90. Mean scores on the performance scale were 80 and 83 respectively. Verbal and performance scores are provided in Table 7.2. Performance subtests were slightly lower than verbal in both groups (Table 7.2). Both groups had below-normal full-scale IQ. On average, subjects with right hemisphere lesions performed slightly better than those with left hemisphere lesions. This difference appeared to be due to greater lesion size in the left hemisphere damaged group.

Subjects who acquired a unilateral cerebral lesion as adults have a very different pattern of cortical function: patients with right hemisphere lesions

Table 7.2.   IQ after unilateral brain injury at various ages. VIQ = Verbal IQ, PIQ = Performance IQ on the Wechsler Scale.

| Right hemispherectomy | | | Left hemispherectomy[a] | | | Right focal lesion | | | Left focal lesion[b] | | |
|---|---|---|---|---|---|---|---|---|---|---|---|
| *Infant cases (< 12 months at onset)* | | | | | | *Perinatal onset* | | | | | |
| VIQ | PIQ | N | VIQ | PIQ | N | VIQ | PIQ | N | VIQ | PIQ | N |
| 80 | 78 | 11 | 87 | 82 | 13 | 90 | 83 | 6 | 82 | 80 | 19 |
| *Childhood cases (1–13 years at onset)* | | | | | | *Postnatal onset (>1 month–10 years at onset)* | | | | | |
| VIQ | PIQ | N | VIQ | PIQ | N | VIQ | PIQ | N | VIQ | PIQ | N |
| 89 | 69 | 4 | 74 | 69 | 7 | 83 | 83 | 8 | 74 | 74 | 6 |
| *Adult cases* | | | | | | | | | | | |
| VIQ | PIQ | N | VIQ | PIQ | N | | | | | | |
| 96 | 68 | 3 | — | — | 0 | | | | | | |

a. Hemispherectomy data are from St. James-Roberts, 1981.
b. Focal lesion data are from Levine, Huttenlocher, Banich, and Duda, 1987.

have decreased performance IQ and preserved verbal IQ, while the opposite is the case in those with left hemisphere lesions. Decrease in IQ has been reported to occur after left perinatal lesions, but not in subjects with left-sided lesions acquired after the age of 1 year (Riva and Cazzaniga, 1986).

The IQ effects of early unilateral brain damage can be observed in patients with hemispherectomy (St. James-Roberts, 1981). These individuals represent the extreme case in which all or nearly all cortical functions are represented in a single cerebral hemisphere. However, even after hemispherectomy there may be residual functional activity, detectable by PET, in remnants of the removed hemisphere (Muller, Chugani, et al., 1998).

While there have been a few patients with tested IQ after hemispherectomy in the superior range (Smith and Sugar, 1975), on average there is a decrease in mean IQ of about 20 points below the normal mean (St. James-Roberts 1981; see Table 7.2). This does not appear to be due to damage to the remaining hemisphere in that, in those patients with partial lesions of a hemisphere, reduction in IQ is proportional to lesion size (see above). In other words, the data suggest that IQ in individuals with early-acquired focal brain lesions is related to the total amount of cortical volume that is available for information processing. The available hemispherectomy data, summarized by St. James-Roberts, show an average verbal IQ of 87 in left hemispherectomy cases and 80 in right hemispherectomy cases, and a performance IQ of 82 and 78, respectively. The somewhat higher mean scores in the left hemispherectomy group appeared to be due to the inclusion of a single outlier with high IQ (Smith and Sugar, 1975).

The IQ changes after focal cortical lesions in infancy can be explained as a "crowding effect" in the remaining normal hemisphere. More functions have to become represented in the normal hemisphere. This may make it necessary for one cortical region, normally specified for processing of one specific type of information, to subserve multiple functions. There may be competition between different functional systems for cortical representation. Some, such as language functions, appear to compete better than others, such as visual-spatial (perceptual) functions. This is suggested by studies of children with perinatal unilateral brain damage, which show constructional apraxias (for example, the inability to copy complex shapes) and deficits in the perception of shapes, similar to what is seen in adults with nondominant parietal lobe lesions (Kohn and Dennis, 1974; Stiles and Nass, 1991). A competitive edge of verbal over spatial functions is also suggested by the IQ scores of persons with perinatal unilateral brain lesions, including those with hemispherectomy, which show higher verbal than performance IQ scores irrespective of the side of damage (see above).

It has also been found that functions that are learned late develop less well than earlier ones. Children with early-acquired unilateral brain lesions appear to have disproportionate difficulty in learning to read (Banich et al., 1990). There also is an age-related decline in IQ in children and adolescents with fetal or infantile onset brain lesions, which suggests that later-acquired reasoning and other higher cortical functions may be more impaired than more basic functions such as language (Banich et al., 1990). What emerges is a remarkable recovery of some functions, especially those related to language, and relative impairment of others, even when the lesion is located outside the usual area of cortical representation of that function. Plasticity in the immature brain does not come without a price.

### Plasticity in Prefrontal Lobe Functions in Subhuman Primates

Evidence of considerable lesion-related plasticity in both the anatomy and function of the prefrontal cortex has been found in rhesus monkeys. This has been seen most prominently in animals whose lesions were made during fetal life. Goldman and Galkin (1978) made lesions in dorsolateral prefrontal cortex in fetal monkeys, with subsequent replacement of the fetus into the uterus. Unlike animals with later lesions, the monkeys with a history of fetal surgery had evidence of marked anatomical reorganization of the prefrontal cortex. Both in animals with unilateral lesions and in animals with bilateral lesions there was formation of new (heterotopic) gyri that histologically resembled normal prefrontal cortex. The parvocellular portion of the mediodorsal thalamic nucleus, a major input to dorsolateral prefrontal cortex, did not show the degeneration seen after postnatal lesions. After unilateral prefrontal fetal lesions there also was evidence of marked reorganization of corticofugal pathways on the unoperated on side, with enlargement of projection to the contralateral caudate nucleus (Goldman, 1978). Careful testing of prefrontal lobe functions showed that —unlike postnatally lesioned monkeys—the ones with fetal lesions functioned at a level comparable to that of normal monkeys. The findings indicate a remarkable and probably unique ability of prefrontal lobes to recover from fetal injury.

A very different outcome was reported when the effects of fetal versus neonatal lesions were compared in felines (Villablanca, Hovda, Jackson, and Gayek, 1993; Villablanca, Hovda, Jackson, and Infante, 1993). The cats with lesions made during the last trimester of pregnancy functioned less well than those with neonatal lesions on a large range of motor and visual tasks. These animals had widespread, bilateral dysgenesis of the cerebral cortex, suggesting a diffuse effect of fetal neurosurgery on brain development in this species.

### Summary

The available evidence indicates an interplay of genetic and environmental influences on the anatomical, physiologic, and functional development of the cerebral cortex. Genetic influences predominate in cortical regions that mediate motor and primary sensory functions. Environmental influences are more marked in areas that subserve higher cortical functions. Examples of this can be found in reading, development of which is entirely dependent on training, and in musical competence, which requires a high level of training, starting early in life. However, even in these functions, genetic influences can be discerned. These include the fact that specific gene defects are associated with decreased ability to learn to read (dyslexia) (Pennington, 1991), and that musicality often appears to have familial occurrence. A specific genetically determined anatomical substrate is likely to be necessary for the optimum expression of cognitive functions. Most likely, the genetic programs provide the basic outline of cellular organization and connection, the substrate on which environmental influences can exert their effects. Environmental factors, in turn, appear to determine many individual differences, including differences in knowledge and motivation.

# 8

## ADULT PLASTICITY

> It may be useful to question the simplistic view that the brain becomes unbendable and increasingly difficult to modify beyond the first few years of life. Although clearly much of brain development occurs late in gestation through the first years of postnatal life, the brain is far from set in its trajectory, even at the completion of adolescence.
>
> Charles A. Nelson and Floyd E. Bloom, 1997

The ability of the brain to mold itself in response to changed environmental conditions does not end with puberty. The brain, and especially the human brain, undergoes important changes in its organization throughout the life span. We can learn at any age—although this may become more difficult with advancing years. The available evidence also suggests that the strategies that the brain uses during learning may differ to some extent depending on age.

### Mechanisms of Adult Plasticity

Plasticity in the adult brain has been the subject of intense recent study, both in humans and in animals. Human studies have largely focused on lesion-related plasticity, such as the recovery from stroke and other acute focal brain lesions. However, experimental learning paradigms and changes induced by learning in the adult brain are also beginning to be explored in humans. Several cellular mechanisms by which the adult brain can adjust to changes in the environment or in sensory input have been defined, including the following.

## DECREASE IN INHIBITION

Many connections between the cortex and the periphery as well as intracortical connections are not normally apparent because of inhibitory influences (Wall, 1971). These have been demonstrated most clearly in sensory systems, but undoubtedly occur throughout the brain. For example, in the retina, a stimulus, such as a point of light, produces an increase in the discharge rate of ganglion cells (the cells that send retinal information to the visual cortex) near the center of the stimulus, but inhibits the activity of neighboring ganglion cells (Kuffler, 1953). The presence of an excitatory center and an inhibitory surround acts to amplify the message that is sent to the cerebral cortex. Similar inhibition of neuronal activity at the periphery of an area of increased activity occurs in the cerebral cortex. It has been studied especially well in the somatosensory cortex, where sensory stimulation of a point on the skin excites neurons near the center of the area of cortical representation and inhibits activity in neurons near the fringe. This makes the receptive field appear smaller than its actual size. The inhibition is due to activation of inhibitory interneurons near the center. Destruction of these cells in a focal cerebral lesion will remove the inhibition in the periphery, and the inhibitory fringe may become a new center of activation. In this manner, a focal lesion of the somatosensory cortex can show rapid functional recovery, without creation of new neuronal circuits (Mountcastle and Darian-Smith, 1968; Merzenich et al., 1984).

## INCREASE IN SYNAPTIC STRENGTH

It is now clearly established that the effectiveness of synaptic connections is modulated in response to functional demands. Synaptic transmission becomes facilitated in a pathway that is frequently activated. In this way, repeated practice of a task leads to increased speed and accuracy of performance. Increase in synaptic strength may well be the major mechanism for learning and for recovery from brain injury in the adult.

Activity-related changes in the excitability of synapses have been studied most extensively in hippocampal neurons. Repetitive stimulation leads to an increase in the excitability and facilitation of transmission in hippocampal synapses. These effects persist for some time after the initial stimulus, and subsequently show gradual declines (long-term potentiation, LTP). They appear to be mediated by modification of calcium channels in the neuronal membrane that leads to an influx of calcium, which in turn increases neuronal excitability. LTP is probably the major mechanism by which learning and the initial steps of memory consolidation take place in the brain, at any age (Kandel, Schwartz, and Jessell, 1991). Learning and forgetting therefore imply a change in the synaptic organization of the brain: increase in synaptic strength in pathways that are utilized and decrease in excitability in circuits that are fallow. The ability to learn is one of the major manifestations of neural plasticity, which persists throughout the life span.

### Axonal Sprouting

The sprouting and elongation of new axons (axon collaterals) in response to focal injury appears to occur at all levels of the nervous system. Preserved neurons at the edge of a lesion send new axonal branches (axon collaterals) into the damaged region and innervate dendrites that have lost their synaptic input. Axonal sprouting leads to new synapse formation at the point of contact of axonal sprouts with these dendritic trees. This mechanism for recovery from a focal lesion has been implicated in the reaction of the somatosensory cortex to loss of its input from the skin, both in animals (Merzenich et al., 1984) and in humans (Pons et al., 1991) (see Chapter 4).

Axonal sprouting in response to loss of input has also been studied in the primary visual cortex. Darian-Smith and Gilbert (1994) reported axonal sprouting in neurons of the visual cortex after placement of bilateral, symmetrical retinal lesions. At first, the cortical area deprived of input from the retina is electrically silent. However, after several weeks this cortical region begins to respond to visual stimuli from neighboring, uninjured retinal areas. Labeling of axons by antegrade transport has shown that this is accompanied by the formation of laterally projecting axonal sprouts of cortical neurons near the margin of the area that has lost its input. This results in enlargement of the receptive fields of retinal areas near the lesion.

A well-known example of axonal sprouting occurs in skeletal muscle after partial denervation. Partial loss of anterior horn cells in the spinal cord at first leads to denervation of the muscle cells that had been innervated by these neurons. Neighboring surviving anterior horn cells then send axonal sprouts to reinnervate these muscle cells. This leads to recovery of function, provided that the loss of anterior horn cells remains incomplete (Figure 8.1).

## FORMATION OF NEW SYNAPSES

The formation of new synapses has been demonstrated to occur in animals in response to enriched environmental input (Turner and Greenough, 1985; Kleim et al., 1996). If this is a large factor, it is likely to be balanced by loss of other synaptic connections, since there is no evidence for any large net increase in synapses in the cerebral cortex during the adult years. Protein synthesis and formation of new synapses appear to underlie long-term memory consolidation. The formation of new synaptic connections is supported by anatomical studies with the Golgi method, which show dendritic sprouting and spine formation in response to increased functional demands in both animal and human cerebral cortex at all ages.

## FORMATION OF NEW NEURONS

Recently, there has been much interest in the question of whether new neurons form during adult life in the human forebrain, and whether these may be utilized for learning and memory. It is now well established that new neurons form in adult primate brains in specific areas, especially the dentate gyrus in the hippocampus. This has even been shown to occur in the adult human brain (Eriksson et al., 1998). Evidence for adult neurogenesis in cortical association areas, but not in the primary visual cortex, has been reported in subhuman primates (Gould, Reeves, et al., 1999). These cells appear to originate in remnants of the periventricular germinal matrix and to migrate to the cerebral cortex. The number of such cells appears to be small. It is as yet unknown whether they survive and whether they can be incorporated into a functioning system. As yet, there is no direct evidence for widespread adult neurogenesis in the cerebral cortex of humans. Nevertheless, the findings in subhuman primates are of great interest. Methods may be discovered by which the number of such cells could be increased in the case of focal brain injury, and by which newly formed neurons could be directed to the site of the lesion, contributing to reparative processes.

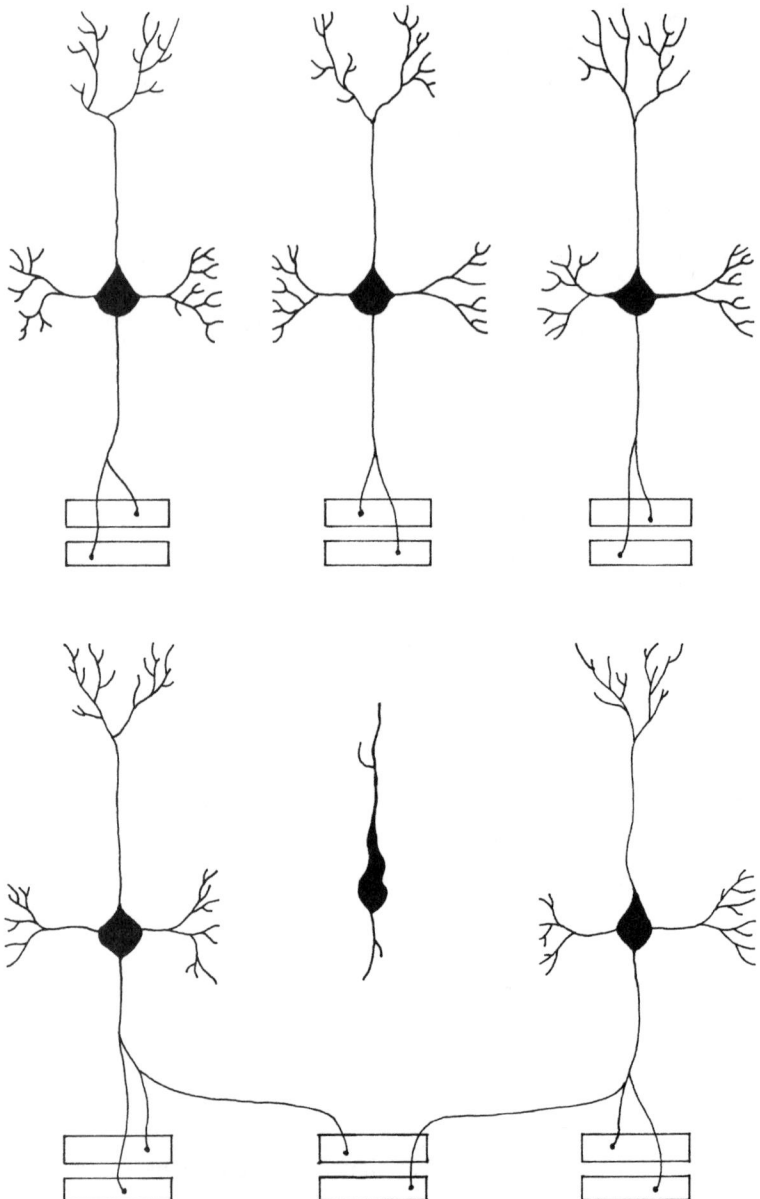

Figure 8.1. Axonal sprouting in response to neuronal death. The death of a neuron results in the innervation of its target by neighboring healthy neurons. The upper half of the figure represents three schematic pyramidal neurons with axons that innervate separate targets (represented by rectangles). The lower half of the figure represents death of the middle neuron and innervation of its former targets by axon collaterals (sprouts) from the two remaining healthy neurons. (Drawing by Arun S. Dabholkar.)

## Dendritic Sprouting

Dendritic sprouting appears to be a compensatory reaction to the loss of nearby neurons that have a similar function. It has been demonstrated in aging human cerebral cortex, where it appears to be a response to normally occurring neuronal loss (Buell and Coleman, 1981).

### Comparison of Infancy and Early Childhood Plasticity with Adult Plasticity

Greenough, Black, and Wallace (1987) distinguish the "experience expectant" plasticity that is seen in the infant and young child from the "experience dependent" plasticity seen in the older child, adolescent, and adult. "Experience expectant" plasticity refers to the readiness of the immature brain to respond to and to become shaped by the normal human environment. It is important for the development of basic, "obligatory" brain functions, of those functions that develop in every normal child, including language comprehension and speech. By the time the child is 1 year old, a huge number of synaptic contacts are present in the cerebral cortex. Most of these appear to be as yet unspecified for function. At this point external input (experience) is necessary for inclusion of some of these synapses into functioning circuits, while others remain inactive and eventually disappear. Whether any given function will develop or not, and how it will develop, depends to a large extent on the type and timing of environmental input. For example, in language areas of the cerebral cortex, the brain of the 1–2-year-old has nearly reached the maximum number of synaptic contacts and is ready to begin to process language. The type of language (English, Japanese, or even sign language) depends on language input during the early childhood years. Lack of exposure to any language during infancy and childhood, as in so-called feral children, appears to permanently impair language development (Curtiss, 1978; Lane, 1976). The best studied of such children, named Genie, had almost no exposure to language until the age of 13 and 1/2 years. She was mute when she was discovered at that age, and subsequently developed language imperfectly, with deficits in speech (expressive language) rather than comprehension. However, even with limited exposure, basic vocabulary and grammar will emerge.

The age at the time of exposure appears to be important in the acquisition of obligatory cortical functions. By the adult years, basic skills such as oral language have usually been learned, if they are to develop at all. Learning of these skills as an adult is still possible, but appears to be more difficult. This may be because in the adult some mechanisms for anatomical plasticity are no longer in place. Environmental changes can no longer modify cortical circuits by the utilization of functionally unspecified synapses. This pathway for the formation of new functional connections appears to be exhausted by late adolescence. The decline of the pool of labile synapses is gradual, extending over several years. There certainly is no anatomical evidence for early critical periods for learning or for recovery from focal brain injury that end abruptly and completely in late infancy (Bruer, 1999).

"Experience dependent" plasticity refers to the ability of the *adult* brain to adjust itself to changes in environmental conditions. It relates to the learning of special skills that may be mastered by only a few, and that require specific training. Unlike language learning in the infant, which proceeds without effort, it requires motivation and concentration on the task. Learning to fly an airplane or learning calculus are examples of tasks that require "experience dependent" plasticity of the brain. These are skills that are usually learned in the adolescent or adult years. There is, however, no clear dichotomy between "experience expectant" plasticity, as confined to the child, and "experience dependent" plasticity, as seen only in the adult. Older infants and children are already capable of engaging in much experience dependent learning. Children, as well as adults, may learn special skills that are dependent on intense training, and that fail to develop at all without such training. Examples include gymnastics and playing a musical instrument.

Some skills have the possibility of both experience expectant and experience dependent learning. An example would be learning to read. By the age of 6 or 7 years, the child's brain appears to be "reading ready." Reading skills are acquired relatively effortlessly by most children at about this age. The anatomic substrate necessary for reading appears to be especially fertile in the young child. However, a person who was never taught to read at that age can still acquire the ability to read through intensive training as an adult, making use of the experience dependent plasticity of the adult brain.

The existence of adult plasticity makes it imperative to compare the effects of environmental changes and of focal brain lesions in the infant with those of similar changes later in childhood and in the adult. Without such comparisons, no conclusions can be drawn concerning the enhanced plasticity of the developing brain. Unfortunately, adult cases have been included infrequently in studies on neural plasticity in humans. In the case of lesion-related plasticity, it has been difficult to match infancy and adult cases with respect to size and location of the lesions (see below). Hemispherectomy cases avoid the problem of matching for lesion size or location, but very few adult cases have been available for study, and none with left hemidecortication. Comparison of lesion-related plasticity in infancy and adulthood therefore is largely dependent on animal studies. The cat has been an especially good model, since its cortical development at birth approximates that seen in the human. Problems that may be encountered when adult controls are inadequate have already been discussed with reference to the effects of neonatal motor cortex lesions in monkeys (Kennard principle, Chapter 5).

### The Ability of the Adult Brain to Adapt to Environmental Change

Changes in the adult rat brain in response to environmental enrichment have been studied extensively by Greenough and associates (Volkmar and Greenough, 1972; Floeter and Greenough, 1979; Green et al., 1983; Turner and Greenough, 1985; Greenough, Black, and Wallace, 1987; Kleim et al., 1996). The effects depend on the type of environmental change. Proliferation of cortical blood vessels and/or increase in dendritic arborization and in synaptic density have been noted in adult rats on enrichment programs. Depending on the type of enrichment, different structures in the cerebral cortex increase in density or volume. An environment that provides primarily for increased motor activity leads to an increase in the density of small blood vessels in the cerebral cortex. This is probably in response to increased cerebral blood flow to brain regions with high neuronal activity, which require an increase in metabolic substrates, including oxygen and glucose. On the other hand, a complex environment that promotes learning of new motor skills, but not necessarily increased motor activity, is associated with de novo synaptogenesis in the motor cortex (Kleim et al., 1996). Studies utilizing the Golgi method have shown increased numbers and length of dendritic branching in cortical neurons in response to housing in a complex environment (Uylings et al., 1978; Green et al., 1983). Increased thickness of the cerebral cortex and increased number of glia have been noted by M. C. Diamond and associates (1966) in rodents maintained in enriched environments.

In all of these studies, the type of control group that is used is of importance. Controls often have consisted of rats maintained in small, crowded cages. This raises the question whether one is studying hypertrophy due to increased stimulation of the experimental group or atrophy related to deprivation in the "control" group. In human terms, the environment of the enriched groups might be comparable to normal living conditions, while that of the controls would be comparable to being permanently locked into a small cell, with deprivation of most of the normal environmental input.

### Einstein's Brain

In the human, there is some evidence that larger brain volume is weakly correlated with higher intelligence ($r = .35$) (Wickett et al., 2000). However, there is no evidence that the brain as a whole is larger in those exposed to especially challenging environments or in individuals with high creativity and high achievement than in those leading an average existence. The brain weights of prominent scientists and writers have covered the entire range from low to high normal. The brain of the Russian writer Ivan Turgenev weighed 2,012 grams, that of Anatole France 1,017 grams.

The brain of Albert Einstein has been studied most thoroughly. The total brain weight was 1,230 grams, in the low normal range (the mean for adult males of similar age = 1,342 grams). An increased ratio of astroglia to neurons was found in the left parietal lobe (Diamond, Scheibel, Murphy, and Harvey, 1985). We do not know whether this change was present when Einstein proposed his Theory of Relativity as a young man or whether the glial increase represented effects of aging unrelated to his superior mental capacity. However, it certainly is possible that this finding does relate to Einstein's superior mathematical abilities. The left posterior inferior parietal lobe is known to be related to the processing of mathematical reasoning. Lesions in this area are associated with acalculia (inability to calculate). The increased ratio of astroglia to neurons in this region in Einstein's brain may be related to decreased neuronal density secondary to enlargement of the dendritic trees of cortical neurons. Such an effect has been seen in animal studies of environmental enrichment (see above). The decreased neuronal density should be accompanied by increased volume of the same cortical region, if it is related to a higher level of dendritic development. This would be necessary to accommodate a normal number of cortical modules. Recent morphometric studies of Einstein's brain do indeed show hypertrophy of the left inferior parietal lobule (the upper bank of the Sylvian fissure and the region just posterior to the Sylvian fissure), the area that has been implicated in mathematical thought (Witelson, Kigar, and Harvey, 1999). This region had also been noted to be prominent in other mathematicians, including Karl Friedrich Gauss, one of the inventors of calculus (Spitzka, 1907).

These observations again raise the question, already broached in the discussion of musical competence in Chapter 6, whether excellence in specific higher cortical functions may have an anatomical basis, that is, focal enlargement of the cortical areas underlying the specific function. The findings in mathematical geniuses need to be confirmed by the study of larger numbers of subjects. We no longer need to wait for the death of a scientist to study his or her brain. This can now be accomplished in vivo by use of quantitative brain MRI. If the inferior parietal lobule is indeed larger in mathematical geniuses than in people with average mathematical talent, then another question arises. We then need to ask whether the mathematical geniuses are born with an unusually large inferior parietal gyrus or whether this gyrus enlarges in response to extensive use during childhood, adolescence, or young adult life. A known example of the former alternative would be the larger size of the left planum temporale, already present at birth in most humans, which appears to make the left temporal lobe superior to the right for language processing. An example of the latter alternative would be the increased thickness of cortex, increased synaptic density, and sprouting of axons and dendrites that appear to occur in animals exposed to enriched environments. This question could be answered now by in vivo quantitative MRI in infancy and by testing of the correlation—or lack of such—between the size of the critical brain region in infancy and later mathematical ability.

If large-scale formation of new synapses and axonal and dendritic sprouting occur in some systems of the human adult brain, they are likely to be approximately balanced by a decrease in the number and complexity of synapses, dendrites, and axons in other cortical systems. This would be necessary to maintain relatively constant total brain size. As indicated above, there is no evidence that human genius is correlated with overall large brain size. Interestingly, the morphometric study of Einstein's brain provides some evidence concerning this question as well. A region in the prefrontal cortex was found to have an unusually thin cortical ribbon and increased density of cortical neurons, suggesting underdevelopment of this area (Anderson and Harvey, 1996; Hines, 1998).

### Dendritic Sprouting in Response to Neuronal Loss

There is evidence for growth and sprouting of dendrites in human adults under certain conditions. It has been reported that sprouting of dendrites, resulting in greater complexity of dendritic trees of cortical neurons, occurs in normal aging in relation to neuronal loss. The loss of cortical neurons during apparently normal aging has been demonstrated in several studies (Huttenlocher, 1979). Buell and Coleman (1981) studied autopsy tissue of the cerebral cortex in the aged by the Golgi method. They found that neuronal loss appeared to be compensated for by increase in the length and complexity of dendrites on cortical neurons. Presumably, the functional capacity of the remaining neurons is increased. New formation of synapses is likely to occur along the course of elongated dendrites, since there is normally a direct relationship between dendritic length and synapse number (see Chapter 1). In other words, synapses appear to have an approximately equal distribution and fixed density along dendrites. Dendritic sprouting has not been observed in brains from patients with senile dementia, which suggests that the presence of sprouting reflects a functionally important compensatory mechanism. Sprouting of dendritic branches and new synapse formation appear to be common responses not only to cell loss but also to increased functional demand, such as exposure to conditions requiring more complex visual processing (Greenough and Volkmar, 1973) or increased complexity of motor activity (Kleim et al., 1996).

### Changes in Adult Human Cerebral Cortex in Response to Training

Changes in the cerebral cortex analogous to those reported after motor training in animals have recently been found in adult humans by the fMRI technique. Karni and colleagues (1995) had subjects undergo daily practice sessions of a complex motor task (rapid sequences of finger movements). The speed and accuracy of performance gradually increased over a period of 4 weeks. At the end of that period, performance of the learned task during fMRI scanning activated a larger area of primary motor cortex (M1) than had been activated prior to the practice sessions. A comparable unpracticed task did not produce increased activation, nor was there increased activation of the motor cortex ipsilateral to the trained hand. The effect persisted for several months, during retention of the learned skill. The results are consistent with the establishment of new synaptic circuits, either by utilization of circuits previously dedicated to other motor tasks or by dendritic sprouting, as was found in the animal experiments.

A change in structure of the adult brain related to acquisition of an unusual skill has been reported by Maguire and colleagues (2000). These authors obtained structural MRI scans of a group of London taxi drivers. They found that the posterior hippocampus (an area concerned with the processing of spatial memory) was increased in volume in the taxi drivers. In the right posterior hippocampus, the increase was greatest in taxi drivers with the longest periods on the job, a finding presumably related to greatest proficiency in spatial navigation. The enlargement of the posterior hippocampus was associated with a decrease in size of the anterior hippocampal formation. The findings mesh well with reports that neurogenesis in the dentate gyrus persists in the adult human brain (Eriksson et al., 1998). Enlargement of the hippocampus in the adult therefore could be secondary to increased neuronal number as well as to increased synaptic density or dendritic complexity. The findings also are in agreement with fMRI studies that show activation of the hippocampus during navigation and spatial memory tasks in humans (Maguire, Burgess, and O'Keefe, 1999), and with clinical studies that show spatial memory deficits in patients with right hippocampal damage (Abrahams et al., 1997), which all point to a role for the hippocampus in spatial memory functions. However, confirmation of the findings in the taxi drivers is needed. A control group that is derived from the archives of an MRI scanning facility, as was used in the London study, may well differ significantly from a group of prospective taxi drivers, prior to the institution of any training. For example, prospective taxi drivers may be preselected for high spatial ability, and hence may have larger hippocampal volumes prior to training than the type of controls used in the study. A definitive study would need to be prospective; that is to say, hippocampal volume would need to be determined before and after navigational training in the same subjects. As a minimum, a control group of prospective taxi drivers, prior to the start of training, would have to be included. Regardless of the cause of the change, the findings do suggest a significant increase in size of a specific cortical region in relation to a special skill, reminiscent of the reported enlargement of a specific brain area in Einstein's brain, discussed earlier in this chapter.

### Adult Plasticity in the Somatosensory System: Effects of Changes in the Periphery

The persistence of plasticity into the adult years is especially prominent in the somatosensory cortex. Merzenich (1984), in a now classic experiment, removed a single digit from the forelimb of monkeys and studied the map of the cortical representation in the hand area in the somatosensory cortex of the opposite hemisphere. At first, cortical neurons in the region of the somatosensory cortex that had lost its peripheral input could not be activated by any form of somatosensory stimulus. However, within 2 months these neurons regained their response to sensory stimuli. The receptive fields of the two digits neighboring the amputated digit had expanded into the field of the missing digit. The functional significance of this type of change in the cortical map is not known. This may be an instance in which plasticity imparts no significant benefit and may even contribute to a negative effect, such as phantom limb sensations (see below). This type of plasticity persists in the adult, and thereby differs from plasticity in other sensory systems, such as, for example, plasticity in the visual system related to strabismic amblyopia, which has an upper age limit of 5–7 years.

Merzenich and associates did not find evidence of expansion of the receptive fields into the representation of the amputated area when the amputation was more extensive, for example when it involved two adjacent fingers. Reorganization appeared to be limited to the cortex within about 1 millimeter from the cortex that had retained its normal innervation. This suggested sprouting of axon collaterals on thalamocortical afferent fibers as a mechanism for the plasticity of the cortical map (Calford and Tweedale, 1990), since these usually do not extend beyond 1 millimeter. However, more recently, Pons and associates (1991) reported a much more extensive reorganization of somatosensory receptive fields that occurs more slowly, over several years (see also Pons, 1994). In adult macaque monkeys, they found extensive changes in cortical maps years after deafferentation of an upper limb. The cortical region that had lost its normal somatosensory input now responded to stimulation of the contralateral face and contralateral foot. The mechanism for this more extensive, slowly developing plasticity of cortical maps is as yet unknown.

Adult plasticity in the somatosensory system has also been studied in humans, as was already discussed in Chapter 3. Examples are phantom limb sensations and their modification. These sensations can be reproduced by sensory stimuli applied to the skin neighboring the amputated limb, including the ipsilateral face after amputation of an upper extremity. This suggests that, in humans as in subhuman primates, the area of sensory representation of the neighboring skin regions has extended into the cortical regions that normally represent the amputated areas (Ramachandran, 1993; Julesz and Kovacs, 1995).

### Plasticity and the Recovery from Focal Brain Lesions in the Human Adult

Recovery, both spontaneous and in response to training, has been studied extensively in adult stroke patients. There often is considerable spontaneous recovery. This may be due less to relocation of function in undamaged brain regions, such as areas in the opposite hemisphere, than to recovery of function near the border of the infarct. In these border zones, there may be transient neuronal dysfunction related to brain edema and excitotoxicity, especially the excessive release of the excitatory neurotransmitter glutamate from damaged neurons (Rothman and Olney, 1986).

One recent study suggests that aphasia in adults fails to recover or recovers poorly if Wernicke's area is destroyed (Heiss et al., 1999). The adult cortex is better able to reorganize speech functions with lesions in other language-related areas (Broca's area, the arcuate fasciculus). Evidence of activation of the nondominant hemisphere was seen by PET scan in some cases, but was associated with poor language recovery. The findings suggest a clear difference between adult and childhood strokes. In childhood strokes, the area of focal damage—Wernicke's area, the arcuate fasciculus, or Broca's area—appears to make little difference. In all types of unilateral lesions involving language areas in children under the age of 7 years, there tends to be mutism, followed by recovery of speech functions. One shortcoming of the adult stroke study by Heiss and colleagues was the short follow-up period (only 8 weeks from stroke to study). Prior studies have shown improvement of aphasia after adult stroke for up to 1 year (Selnes, 1999). The study therefore needs to be extended to include longer post-stroke follow-up.

Some data relevant to the question of whether relocation of function in the opposite (normal) hemisphere occurs in adults are available from patients with large hemispheral gliomas who were treated with hemispherectomy. Only patients with nondominant (right hemisphere) hemispherectomy have been available for study. In three such patients, reviewed by St. James-Roberts (1981), the effects of hemispherectomy on cognitive functions differed from those with early childhood lesions. The adult cases had normal dominant hemisphere (language) functions, and severe nondominant hemisphere (visual-spatial) dysfunction. Verbal IQ was normal, a mean of 96; performance IQ (mainly nondominant hemisphere functions were tested) was 68, in the mentally defective range. In contrast, the subjects with early right hemispherectomy had both verbal and performance IQ scores that were on the average low normal, in the 80s (Chapter 7). These data show little or no evidence of reorganization of nondominant hemisphere functions in the dominant hemisphere after adult hemispherectomy. Similarly, adults with massive strokes that destroy most of the dominant cerebral hemisphere have permanent loss of language functions. Gait also is more severely affected in adults with large strokes than in infants with strokes. Many of the adults have major gait disturbances when tested 6 months after the stroke (G. Kwakkel et al., 1999), including the inability to walk without assistance. The adult stroke patients also tend to have persistent impairments in the activities of daily living (ADLs). In contrast, children with strictly unilateral perinatal brain damage, including those with early hemispherectomy, have good functional gait and are fully independent in daily living.

### The Effects of Training on Adult Stroke

There remains the question whether extensive training can improve the outcome in stroke patients, as far as sensory, motor, and speech functions are concerned. Recent evidence suggests that extensive training may have positive effects on some functions. Earlier randomized, controlled studies of stroke rehabilitation were reviewed by dePedro-Cuesta and colleagues

(1992). There were few studies that showed significant benefit for motor activities, especially activities of daily living in the home. Methods that were evaluated included conventional physiotherapy, biofeedback, and perceptual training. More recent studies suggest that novel approaches to motor rehabilitation may hold more promise (Bach-y-Rita et al., 1981). Taub, Miller, Novack, and their colleagues (1993) studied the effect of restraint of the normal hand on the function of the paretic hand in adult stroke patients, 1–18 years after the stroke, when spontaneous improvement is unlikely. Limb restraint on the normal side for 90 percent of waking hours during 12 days was paired with daily, intensive therapy consisting of practice of simple motor tasks or of a shaping program, in which a desired motor objective was approached in small, sequential steps (Taub and Crego, 1995). Both programs led to significant improvement in paretic hand function, which carried over into the activities of daily living and therefore provided significant benefit to the patients. However, only about 20–25 percent of stroke patients had sufficient baseline ability to move the paretic hand and fingers to be able to benefit from the program.

Speech therapy for aphasia due to stroke in the adult appears to have a benefit for at least several months after the stroke. The effects of intervention on language recovery appear to be greater than those on motor performance. Even time spent practicing language tasks with minimally trained laypersons appears to improve speech functions in adult stroke patients (de Pedro-Questa et al., 1992). The results suggest that lesion-related plasticity is greater for language than for motor functions in the adult as well as in the child.

An interesting effect of training on lesion-related sensory deficit has been reported recently. Kasten, Wust, Behrens, and their colleagues (1998) reported training-related improvement in the visual field in adult patients with unilateral lesions in the central visual pathways. This study utilized a computer-based shaping program, which led to improvement of the visual field at the border between normal and abnormal. The size of the recovered field was modest, but may be of functional significance, for example in the ability to safely operate a motor vehicle.

The available evidence suggests that significant improvement in adult stroke patients can be achieved through rehabilitation programs. Several new programs including dominant limb restraint and shaping methods hold particular promise.

Comparison of plasticity in adults versus infants has been difficult in humans. This has been especially true for lesion-induced plasticity. Matching of pre- and perinatal with adult focal brain lesions for size and location presents a number of problems. The reaction of fetal tissues to injury differs markedly from that of adult tissue. In the fetus, focal damage often leads to complete loss of tissue, which is replaced by a cavity (porencephaly), while in the adult there is attempted tissue repair, with formation of a glial "scar," which merges gradually into normal tissue. The limits of an adult lesion, such as stroke, may therefore be difficult to define. In the infant with porencephaly it may be difficult to determine which brain regions are affected because the presence of a cavity may lead to distortion of the normal gyral patterns. This may occur by extension of remaining brain tissue into the empty space, or by compression of tissue if the fluid in the cavity is under pressure.

In the comparison of functional plasticity in normal infants and children versus adults, a number of factors may enter into the outcome that may be difficult to control. For example, in second language learning, it is generally stated that young children learn faster than adults do. However, this conclusion is not universally accepted. Under certain conditions, adults may learn faster, and when this occurs it is related to factors such as motivation and a greater ability to concentrate on a task.

Many of the difficulties encountered in the comparison of developmental and adult plasticity in humans can be avoided in animal experiments. Careful comparisons of lesion-induced plasticity in infant and adult animals have been made. Goldman and Galkin (1978) made lesions in the prefrontal cortex of fetal monkeys and compared the effects with those of postnatal resections at various ages. The animals with early fetal lesions (6 weeks prior to term) had the most marked evidence of both anatomical reorganization and sparing of prefrontal lobe function (see also Chapter 7, p. 168).

Extensive comparisons of the effects of lesions at various ages were made by Villablanca and associates, with attention given to lesion size and location and to time elapsed between the placement of the lesion and testing for residual function or for anatomical reorganization. These experimenters carried out a number of studies comparing the effects of neonatal versus adult hemispherectomy in cats (Gomez-Pinilla, Villablanca, et al., 1986).

The infancy lesions were made between the ages of 1 and 12 days, a period in which the kittens have the level of development observed in the human neonate. Careful studies of anatomical changes and of motor activity were carried out. Anatomical rearrangements were much more extensive in animals with neonatal hemispherectomy. These animals developed a number of efferent pathways in the remaining intact cerebral hemisphere, which are not present in normal animals. These included a crossed cortico-rubral tract, connecting the motor cortex of the residual cerebral hemisphere with the opposite red nucleus in the mesencephalon (midbrain). A de novo crossed cortico-rubral connection has also been described in monkeys after perinatal unilateral removal of the motor cortex (Sloper et al., 1983).

Cats with hemispherectomy in infancy had better function than those with adult hemispherectomy in a large number of parameters (Burgess and Villablanca, 1986; Burges, Villablanca, and Levine, 1986; Hovda and Villablanca, 1989, 1990). In a more recent study, hemispherectomy was carried out at various ages between postnatal day (P) 10 and adulthood. Shrinkage of the thalamus ipsilateral to the side of hemispherectomy occurred in all animals, but was less severe when hemispherectomy was carried out prior to P60. Significant shrinkage of the thalamus ipsilateral to the remaining hemisphere as well as shrinkage of the neocortex was seen after P30 (Schmanke, Villablanca, Lekht, and Patel 1998). Behavioral testing of these animals showed that the ones with hemispherectomy prior to P60 performed better on a large range of tasks than those with later hemidecortication. An exception was the forepaw placing reaction, which did not recover at any age, a finding reminiscent of the failure of recovery of hand functions in humans with unilateral perinatal brain lesions. The age-related changes had an upper age limit for optimum function between P30 and P60. However, some advantage over the adult was seen up to age P120 (Villablanca, Carlson-Kuhta, Schmanke, and Hovda, 1998). The time window for lesion-related plasticity in cats closes gradually, as is true of other forms of age-related cortical plasticity.

Differences between neonatal and adult hemispherectomy in cats have also been reported in the cerebral metabolic rate in the remaining hemisphere. The adult hemispherectomy animals had significantly lower cerebral metabolic rates for glucose (CMRglc) than either neonatally operated on animals or normal controls. A similar difference between adult and infant hemispherectomy animals was found in the CMRglc of the thalamus and basal ganglia on the side of decortication (Hovda, Villablanca, Chugani, and Phelps, 1996).

The functional superiority of animals with neonatal versus adult unilateral brain lesions, including hemispherectomy, has also been reported in rodents (Hicks and D'Amato, 1970; D'Amato and Hicks, 1978; Kolb and Whishaw, 1989, 1998) and in rabbits (Hobbelen and van Hoff, 1986). There are at present no known examples of greater anatomical or functional plasticity in adults than in the immature organism.

### Summary

The data on adult plasticity suggest that there is considerable residual malleability of the brain related to environmental variations. The data on lesion-related plasticity indicate differences between different systems. Remarkable adult plasticity has been demonstrated in the somatosensory system. However, where developmental versus adult plasticity on a given task has been compared, adult plasticity has not been found to be greater than that in the developing brain, and has usually been less. This may be due to the developmental loss of some neural mechanisms for plasticity. One such mechanism may be the functional specification of unspecified, labile synapses for the construction of new neuronal circuits.

# 9

There is a very important time in a child's life, beginning at birth, when he should be living in an enriched environment—visual, auditory, language, and so on—because that lays the foundation for development later in life.

Torsten Wiesel, 1982

In previous chapters, the remarkable range of ways in which the structure and function of the developing brain can be modified was explored in several cortical systems. Whereas just a few years ago only a few instances of developmental plasticity were known, and these were largely confined to the visual system, now many are described, in multiple functional systems. No longer is it possible to consider instances of malleability as unimportant exceptions to the rule that brain development progresses in a series of immutable, innately determined steps. The examples are too many and are seen universally. They make a good case for largely environmentally regulated rather than innate cortical systems. A correlate to high plasticity of the brain is the finding that the cerebral cortex is less modular than previously thought, and that in many ways it acts like a multipotential distributed system.

Another new development is the realization that the brain has an amazing range of possible strategies, which can be used to implement environmentally determined changes. Some of these, namely those that can mediate environmentally induced changes at any age, were listed in the chapter on adult plasticity. In this chapter, several strategies that are available only in the immature brain are added. So-called windows of opportunity have been described, owing to the fact that some strategies have age limits. Plasticity may have negative as well as positive effects. An example of possible negative effects is represented by competition between different inputs, leading to crowding.

These issues will be taken up in turn in this chapter. Some of the conclusions need to be tentative, since many of the findings are new, and hence data are often limited and not yet confirmed by replication.

### Contributions from Developmental Anatomy

Data derived from the study of the developmental anatomy of the cerebral cortex have contributed significantly to our knowledge of developmental events, and to theories concerning the development of cortical function and of cortical plasticity. The work most relevant to cortical plasticity has been related to the formation of connections between neurons, namely axons, synapses, and dendrites. The anatomical data have provided information about the timing of developmental events. They delineate an early period of rapid formation of neuronal connections, between birth and the age of 2 years in the human. This appears to be associated with the emergence of the basic functions of the different cortical regions. A second period, between the age of 1 year and early adolescence is a time of exuberant connections in the cerebral cortex. In terms of function, it correlates with heightened plasticity, both with regard to response to environmental stimulation and with regard to the reaction of the brain to focal damage. The persistence of some early connections and the disappearance of others appears to be at least in part environmentally regulated. The input to developing cortex, via the sense organs and from other cortical areas, appears to shape the connections that are included in functioning circuits and become stabilized.

Finally, late childhood and adolescence has been identified as a period during which pruning of many synaptic connections occurs, and when the cerebral cortex acquires its adult structure. What functions, if any, were mediated by the synaptic circuits that are eliminated by synaptic pruning? Clinical studies suggest that late childhood may be a time in which diffuse, bilateral representation of some functions, such as language, becomes restricted to one hemisphere. Circuits dedicated for language processing in the right hemisphere in the young child are an example of systems that are likely to be affected by pruning. Late childhood may be a period during which the cerebral cortex shifts from relatively diffuse processing—a distributed system—to a more modular system, with restriction of cortical representation to specific cortical regions.

Whether synaptic pruning affects cortical function in a positive way is as yet unclear. One possibility that has been considered is that elimination of redundant synaptic connections decreases neuronal activity that is irrelevant to the performance of a given task ("background noise"), and thereby increases the accuracy, efficiency, and speed of information processing. This advantage may, however, come at a price, that is, at the expense of decreased neural plasticity.

The anatomical methods have been of great importance for the definition of some of the mechanisms of plasticity. A number of these, namely those that persist throughout the life span and that mediate adult plasticity, have been discussed in Chapter 8. In addition there are several mechanisms for plasticity that are limited to specific age periods.

### Mechanisms for Plasticity in the Immature Cerebral Cortex

The structure that is primarily involved in cortical plasticity is the synapse. It has been found that the synaptic modification that underlies plasticity can occur in a number of different ways. Most of these have been referred to at various points earlier in this book, in relationship to the discussion of specific instances of cortical malleability. Both the developing brain and the adult brain have several mechanisms by which they can respond to focal injury or to changes in input from the environment. The options appear to be greater in the immature brain, where competition for synaptic sites and functional specification of unspecified, including silent, synapses appear to be major mechanisms. The formation of new synapses, the facilitation of excitatory transmission in existing circuits, and release from inhibitory activity appear to be major mechanisms of cortical plasticity in the adult, but occur in the immature brain as well. As far as we know, all of the mechanisms for adult plasticity discussed in Chapter 8 also are available to the immature cerebral cortex. They are listed in Table 9.1, together with mecha-

*Table 9.1.*   Mechanisms for plasticity in the cerebral cortex.

| Mechanisms available only in immature brain | Mechanisms available throughout the life span |
| --- | --- |
| Utilization of unspecified (labile) synapses, including silent synapses | Decrease in inhibition |
| Competition for synaptic sites | Increase in synaptic strength |
| Persistence of normally transient connections | Dendritic sprouting |
|  | Formation of new synapses |
|  | Formation of new neurons |

nisms that appear to be specific for the developing brain and that are discussed here:

1. The availability of unspecified (labile) synapses. This is likely to be a major substrate for plasticity in the immature brain. Massive overproduction of synapses occurs in infancy, and these may be available for incorporation into functioning circuits until late childhood or early adolescence, when pruning of synapses occurs. Some of these unspecified synapses appear to be electrically silent. An interesting example of the possible importance of silent synapses is their occurrence in the developing somatosensory cortex of rodents during formation of thalamo-cortical connections representing the vibrissae. Isaac and colleagues (1997) and Feldman, Nicoll, and Malenka (1999) found that the disappearance of plasticity in the barrel cortex corresponds to the loss of silent synapses, specifically synapses without electrical activity under normal physiologic conditions. Silent synapses may represent some of the "labile" synapses postulated by Changeux and colleagues (1976, 1980), that is, synapses that are not yet incorporated into functioning circuits. Repeated high-frequency electrical stimulation of silent synapses will transform them into functioning units, which indicates that afferent input is important for the development of function, as predicted by the Changeux hypothesis of synapse specification or validation (see below).

2. Competition for synaptic sites. This mode of environmentally induced change in synaptic organization has been demonstrated most clearly in developing visual cortex (Chapter 4). Initial overproduction of synaptic contacts leads to diffuse representation of input from both eyes to the primary visual cortex. There follows a period of synapse specification that is input sensitive. Contacts that have not been incorporated into functioning circuits

are eliminated. Under normal conditions, when input from the two eyes is symmetric, synapse elimination occurs such that adjacent patches of the visual cortex have alternating input from the left and right eye, the ocular dominance columns. Asymmetric input, due to deprivation of formed visual images from one eye, leads to a predominance of connections from the eye that is exposed to normal visual stimuli. Connections from the visually deprived eye to the visual cortex are decreased, which leads to permanent impairment of vision in the deprived eye. Competition for synaptic sites in the visual cortex is age limited and is not seen after the critical age of 6 to 7 years. This effect may be considered an example of the negative aspects of developmental plasticity, since it results in loss of vision in one eye. Competition for synaptic sites may also account for the so-called crowding effects seen after damage to one hemisphere in infancy.

3. The persistence of normally transient fiber tracts. Many of the axons that are formed during early development normally disappear. This developmental loss may affect entire pathways, such as the uncrossed (direct) cortico-spinal tract from the motor cortex to ipsilateral spinal cord in rodents. This direct pathway may persist after early lesions to the opposite motor cortex. Persistent uncrossed cortico-spinal axons may mediate the partial recovery of motor function after unilateral perinatal injuries to the cerebral cortex (Chapter 5).

### Contributions from Functional Imaging and from Evoked Potential Studies

Recently developed functional imaging techniques have had a remarkable impact by leading to the discovery of many instances of cortical plasticity, some totally unsuspected. Perhaps the most astounding is the finding that the primary visual cortex in the congenitally blind becomes retooled for Braille reading, a task that requires somatosensory input (Chapter 4). Apparently, such input is provided by axonal sprouting from somatosensory afferent fibers to the lateral geniculate nucleus (Asanuma and Stanfield, 1990). In the auditory system, areas that normally are specified for the processing of auditory stimuli process visual input in the congenitally deaf. In general, immature cerebral cortex appears to be able to change its function, depending on demands made by the input to the system.

Functional imaging has brought some other surprises. Many functions appear to be less discretely localized than was previously thought. This is true for language tasks, which appear to be processed to a large extent bilaterally, although left hemisphere activation predominates. Within the left hemisphere, activation extends outside the usually accepted limits of Wernicke's area and Broca's area. Perhaps more surprising, the area of activation expands with increasing difficulty of the language task. What this suggests is a dynamic system that adjusts its size to the task at hand rather than a fixed language cortex (Just et al., 1996). Developmental studies directed at the question of whether cortical localization changes with maturation have to consider this effect, by matching the subjects for the difficulty of the task.

### Nature versus Nurture: Innateness (Genetic Determinism) versus Neural and Functional Plasticity

The existence per se of various means by which the environment may modify developing cerebral cortex does not provide proof that developmental influences are important factors in normal brain development. There is a long history of strong support for both environmentalist and innatist views on brain development. The environmentalist position states that the development of functioning systems depends to a large extent on input to the brain via the sense organs and afferent sensory systems. In contrast, the innatist views development of the brain as an unrolling of genetic programs, with little contribution by environmental input.

#### ENVIRONMENTALIST POSITIONS

An extreme environmentalist stand is found in the writings of K. S. Lashley (1929, 1950) and D. O. Hebb (1949). The cerebral cortex of the neonate was considered to be largely unspecified for function. They referred to this property of immature neurons as equipotentiality. Hebb, in his book *The Organization of Behavior*, defines potentiality in a neural system as "the ability of different cells to acquire the same function in behavior." "Equipotentiality" is "the acquisition of the same function by all cells in a system when excited in a given pattern," presumably by stimuli reaching the system from outside. Equipotentiality of the two cerebral hemispheres would imply that the right hemisphere could be removed without the loss of specific functions, because the left hemisphere as a whole would have learned what the right did. The patterns of neuronal activity are established in response to outside stimuli, that is to say, they are environmentally determined. The brain of the neonate is essentially undetermined, a "tabula rasa."

These ideas obtain some support from observations related to the effects of focal pre- or perinatal brain lesions on cortical function. These effects differ from those of similar brain lesions in the adult. The difference is most evident in language functions, which emerge nearly normally after damage to the left hemisphere in infancy, whereas a similar lesion in the adult would be likely to result in permanent impairment of language functions (aphasia). The two hemispheres of the neonate appear to be nearly equipotential for language functions: the right hemisphere seems to have learned most of what the left would do normally.

Examples where one hemisphere "learns" the same as the other exist even in the case of the adult brain. Calford and Tweedale (1990) reported an example in the adult somatosensory system. They showed that a small unilateral peripheral sensory denervation in adult flying foxes leads to expansion of the cortical receptive field for neighboring skin areas, as predicted from the work of Merzenich and colleagues (Chapter 4). Surprisingly, the receptive field of the homotopic region in the other hemisphere mirrored the change. In other words, the second hemisphere learned what the first had done; it copied the revised sensory map. This is an example of remarkable adult plasticity in the somatosensory cortex. It maintains symmetric sensory representation of the two sides in the cerebral cortex, which may be important for the control of symmetric, bilateral motor activity, as in flying. However, equipotentiality does not exist at any age for several primary visual and voluntary motor functions: lesions in the visual cortex produce a permanent unilateral visual field defect (homonymous hemianopsia) and motor cortex lesions are associated with hemiparesis, no matter at what age the lesion occurs. These effects, in lesions incurred in infancy, are similar to, if perhaps not quite as severe as, those of adult lesions. Some evidence of plasticity has been found in even these relatively fixed systems. Examples include the modification of visual field defects produced by training (Kasten, Woods, Behrens, et al., 1998) and improved function of the paretic hand after restraint of the normal side (Taub, Miller, Novack, et al., 1993).

With the growth of knowledge in developmental neurobiology, there has been refinement of some of Lashley's and Hebb's ideas, while others have been discarded. Equipotentiality, in the sense that all cells in a neural system acquire the same function, is not accepted today. However, the notion that

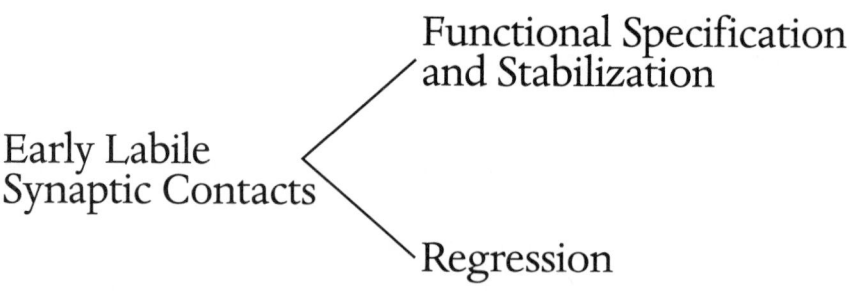

Figure 9.1. Schema for the stabilization of immature synapses by functional specification. (See the text).

many or even most cortical synapses are initially unspecified for function, and that specification occurs in response to external input, has been incorporated in the views of many neuroscientists and developmental biologists. It was clearly stated 25 years ago by the French neurobiologist J. P. Changeux (1976). Changeux viewed immature synapses as being to a large extent functionally unspecified and labile. Stabilization is achieved through the incorporation of these early labile synapses into functional units or circuits. The formation of such circuits is thought to be dependent on external input. Development of the functional organization of a cortical region is thought to depend to a large degree on the input it receives from subcortical and cortical afferent systems, mainly visual, auditory, and somatosensory stimuli in the primary sensory cortex and input from other cortical regions in the association cortex. Changeux proposed that inclusion of a synapse into a functioning network leads to "functional validation." Synapses that are not utilized are eventually resorbed (Figure 9.1).

However, it is also acknowledged that there are limits to the environmental effects and that some cortical systems, such as those underlying basic sensory and motor functions, may develop independently of external stimuli, guided by genetic programs. Clearly, some cortical functions develop without prior experience. An example of this is the visually guided placing reaction of the forelimbs in kittens, which is present in animals in the absence of any prior practice (Hein and Held, 1967; Held and Bauer, 1967). It is proposed that the general outlines of function in a neural system are genetically regulated, but that details and fine-tuning of a system are environmentally determined. Genetic programs may be more important in cortical systems that process basic, obligatory functions than in those used for processing of elective functions, such as the prefrontal and posterior parietal cortex.

G. M. Edelman, in his book *Neural Darwinism* (1987), takes a position that resembles that of Changeux. Edelman stresses the importance of competition between neurons for targets for innervation during development. His work on computer simulation of neural networks provides evidence for the functional organization of initially randomly connected units when exposed to peripheral "environmental" input. This theme is further developed by computer scientists working on the learning of complex functions, including simple language, by computers (Elman and Zipser, 1988; Elman, 1993; Elman, Bates, et al., 1996).

### INNATIST POSITIONS

An innatist view of cortical functions, especially language functions, has been championed by N. Chomsky (1986), J. Fodor (1983), and S. Pinker (1994). This position can be summarized as follows: Neural circuits are laid down early in development, often prenatally. There is early determination of function of a cortical region, based on the genetic program (Crain, 1991). Function develops suddenly, independently of environmental input. This is thought to be true especially for some aspects of grammar that are thought to exist in all human languages (universal grammar).

According to the innatist view, specific cortical functions are represented in circumscribed regions of cortex, or modules, which are present already early in development. Instead of a diffusely organized, largely unspecified cortex the innatist sees fixed cortical specification (Fodor, 1983). Environmental stimulation is thought to have limited effect on the function of the system, although it is acknowledged that marked environmental deprivation may lead to failure of the genetic developmental program.

An innatist stand that includes relatively fixed overall intelligence, or IQ, is found in the work of Herrnstein and Murray (1994). According to this view, higher cortical functions, as measured by intelligence (IQ) or developmental tests (DQ) are thought to develop according to intrinsic programs, and to be affected little by environmental enrichment or deprivation.

The innatist view gets some support from findings in developmental neurobiology. The development of simple motor and visual functions occurs early and very rapidly, consistent with the innatist view. Whole nervous systems are known to exist that function well on the basis of genetically determined developmental programs. A good, extensively studied example is the nervous system of the roundworm *Caenorhabditis elegans*. In this nematode, there is invariant development of a total of 302 nerve cells, most of which are motor neurons. The location of each cell is predetermined and in turn specifies its function and synaptic connections (Sulston et al., 1983). Synapses also occur at specific locations and in specific numbers. This type of nervous system is capable of a number of simple behaviors, including feeding, reproduction, and avoidance of noxious stimuli. The nerve net of *C. elegans* is extremely efficient in its execution of these simple functions. The 302 neurons are programmed to control all functions necessary for survival. However, learning of new skills related to environmental input is limited, if present at all. In this regard, it is of interest to consider the differences between the nervous system of *C. elegans* and the properties of computers that are programmed for learning. Computers capable of learning require an initially randomly connected system, rather than one with preprogrammed functioning circuits.

The higher cortical functions in mammals, and in particular in humans, are bought at the expense of immense enlargement of the nervous system, from a few hundred neurons in *C. elegans* to many millions in primate cerebral cortex. The genetic code is not large enough for specification of all the connections in such a vast system. Nor would a genetically determined cerebral cortex of the size of a human function like a human brain. It would be unlikely to have the built-in capacity to react to millions of often unpredictable stimuli, to plan ahead, and to learn complex new material.

Innatist views derive some support from findings in developmental anatomy. Much of early neurogenesis appears to occur independently of the environment, even in primates. This includes the early production of synaptic connections, including the formation of exuberant connections in some species in at least some cortical areas. There is good evidence that in the visual cortex of the rhesus monkey synaptogenesis is not influenced by visual input (Bourgeois, Jastreboff, and Rakic, 1989). A view that the inexperienced, neonatal brain already is organized into functional modules is tenable for simple functions of the motor and visual cortex. There also is convincing evidence of the importance of genetic factors in higher cortical functions. This evidence derives largely from findings in abnormal populations, where both specific learning disabilities, such as dyslexia (Pennington, 1991; Wolff et al., 1995; Grigorenko et al., 1997; Smith, Penny, and Brower,

1998), and global impairment in IQ, such as in the fragile X syndrome, may be associated with specific gene defects.

As discussed in Chapter 6, equipotentiality has some limits even for language, which is the most plastic cortical function studied so far: right hemisphere language is not quite as good as left, no matter how early in development a lesion has occurred in the left hemisphere. Verbal fluency and the more complex aspects of grammar appear to be selectively impaired. These deficits are relatively minor and should not make us forget the remarkable degree of organization of language that occurs in the right hemisphere when the left is damaged prior to late childhood.

Subtle, probably genetically based, differences in the anatomy of the two hemispheres give the left an edge for language processing. In almost all human brains, the temporal lobe areas on the left that have to do with language processing are somewhat larger than the corresponding areas on the right, providing more anatomical substrate for language processing in the dominant cerebral hemisphere. This difference is present already at birth, several months before the start of language development (Witelson and Pallie, 1973; Chi, Dooling, and Gilles, 1977; Best, 1988). It therefore is unlikely to be secondary to increased functional activity on the left, and most likely represents a genetically determined structural difference.

The innatist view, like the environmentalist one, can explain only some aspects of cerebral cortical development in humans. An extreme innatist view soon runs into difficulties. When it comes to synaptogenesis, there simply are not sufficient genes in the human genome to specify where the billions of synaptic connections that are made in the human cerebral cortex are located. At this point, there is no adequate innatist theory that can explain the emergence of function in human cerebral cortex.

Other problems exist with an extreme innatist position. The question of whether the mind is modular can now be tested experimentally with PET, ERP, and fMRI. This must be considered one of the major contributions of functional imaging to our understanding of brain development. The data, especially from studies of young children, tend to support diffuse rather than modular cortical information processing. For example, language appears to be diffusely and bilaterally represented in the young child. The brain of the infant and young child shows not only bilateral activation during language tasks, but also larger regions of activation on the left side. Areas of activation include the classic speech areas, but extend beyond them, as for instance into the region of the angular gyrus in the posterior-lateral parietal lobe, and into the prefrontal cortex anterior to the classical Broca's area. Representation becomes more "modular" during maturation, but the restriction of representation is gradual. There therefore is plenty of time for the environment to exert its effects.

Another important discovery of functional imaging is that even in the adult the "language areas" of the brain extend considerably beyond the limits suggested by classical neuropathological studies. This appears to be especially so in women, where the right as well as the left hemisphere is activated during reading tasks, as was reported by B. A. Shaywitz and colleagues (1995). The difference between males and females in the cortical representation of cognitive functions may be hormonally regulated (S. E. Shaywitz et al., 1999). The findings suggest that different patterns of cortical organization may be utilized for the processing of the same information by different subjects.

There are now many examples of variations in cortical localization in response to changes in sensory input, if these changes occur during infancy. Examples include the organization of brain regions for the processing of sensory input in the congenitally blind and deaf, which differs markedly from the normal organization (Chapter 4), and reorganization of cortical motor and language areas after pre- or perinatal brain lesions (Chapters 5 and 6). In general, there appears to be a gradual shift in development, from a largely equipotential neural net, or "distributed system," to a more modular system. At the same time, plasticity of the brain decreases.

While the adult brain contains what may be called modules, these are less fixed and specific than is implied by the innatist position. A good example relates to the fusiform gyrus in the inferior temporal cortex, which was found to have activation by fMRI criteria during face recognition tasks (Kanwisher et al., 1997). The initial data suggested a module in the temporal lobe specifically dedicated to the recognition of facial features. This view has been challenged by the recent discovery that the fusiform gyrus has increased cerebral blood flow during the analysis of subtle differences between all kinds of shapes, including figures without faces ("greebles"), by subjects trained to recognize different greebles, and the analysis of differences between makes or models of automobiles by persons trained to recognize subtle differences between them ("car nuts") (Gauthier et al., 1997, 1998). The face recognition module therefore becomes a face plus greeble plus make of car recognition module. Very likely, other shape or feature recognition tasks will also be found to activate this cortical region. Before one can conclude that a given brain region is specified for a given function, one has to show that other functions do not also activate the same region. Whether the mind appears modular or not may depend on the care one takes to search for overlapping functions. The fusiform gyrus may be important for recognition of differences between many types of shapes. Is it, then, a "feature or shape recognition area"? Perhaps, but the basic processing that takes place in this area of the cortex is as yet undefined.

The innatist position with respect to language development has been challenged by recent developmental studies, which show marked effects of the mother's speech and of school attendance on the child's vocabulary and grammar (J. Huttenlocher et al., 1991, 1998). As for IQ, recent intervention studies in early childhood have shown significant increase in IQ scores in children exposed to enrichment programs. This is so despite the fact that items on IQ tests are selected for stability of responses over time (Chapter 7). While it is true that these gains in IQ may be lost if the intervention is discontinued and the child is returned to his or her original suboptimal environment (Bruer, 1999), this fact does not bolster the innatist position. It merely supports the continued importance of environmental effects throughout childhood and adolescence.

Animal experiments support the view that the structure and function of the cerebral cortex can be modified by changes in the environment, even in the adult age groups. Experiments in rodents have shown effects of environmental change ("enriched environments") on the complexity of cortical dendritic trees, synaptic density, and cortical function (Floeter and Greenough, 1979; Pysh and Weiss, 1979; Kleim et al., 1996). Learning and memory are thought to be associated with the formation of new synapses, a process that continues throughout the life span (Kandel, Schwartz, and Jessell, 1991).

RESOLUTION OF THE NATURE-NURTURE CONTROVERSY

In summary, recent developmental data, presented in this book, provide much information that helps to resolve the nature-nurture arguments. The data support neither a strong innatist stand nor an exclusively environmentalist one. Both genetic programs and environmental input are important for the normal development of the brain. The complex interactions between the two, not one or the other, determine the developmental outcome.

### The Functional Importance of Diffuse versus Modular Cortical Representation

The available evidence suggests that postnatal development of the cerebral cortex involves a gradual change from diffuse, bilateral to more focal, unilateral cortical representation. This has been demonstrated most extensively for language functions, as summarized earlier in this chapter, but has also been noted during performance of tasks that activate the prefrontal cortex, such as a "go–no go" task. In this task, the subject is asked to inhibit a given response when a specific stimulus (target) is presented. Response inhibition is associated with activation of the anterior cingulate, and of the middle, inferior, and orbital frontal areas in adults, as has been determined by fMRI. Children had a larger volume of activation, especially in dorsal and lateral prefrontal cortex (Casey et al., 1997). The developmental change from diffuse to a more focal, modular cortical representation is likely to have considerable functional importance, both in a positive and in a negative sense:

1. Diffuse representation in the infant is likely to affect the clinical manifestations of focal cortical lesions. Diffuse (bilateral) representation implies that a focal lesion will have less specific deficit (such as aphasia in a lesion involving the classic speech areas) and more global dysfunction (as manifested by a drop in IQ). Specific deficits should appear with lesions acquired after functional representation becomes restricted, as indeed occurs, for example in language functions.

2. Functional localization in the mature brain is likely to increase the efficiency and accuracy of information processing, since there is less interference from other information-processing tasks occupying the same anatomical area. Response latency for many tasks decreases and speed and accuracy increase at about the age when cortical localization of functions increases (Courchesne, 1978). However, myelination of subcortical and intracortical axons also progresses during late childhood, and could be a factor in the observed decreases in reaction time (Paus et al., 1999).

3. Restriction of cortical representation may free up cortical space for late-developing functions, such as reading and higher mathematics. There may, therefore, be a trade-off between diffuse representation with enhanced plasticity on the one hand and functional localization, which makes possible the emergence of highly specialized, late-developing functions.

4. The maintenance of numerous neural connections requires considerable energy expenditure, as is indicated by the high metabolic rate of the immature cerebral cortex. The high energy requirements of the immature brain may make it more susceptible to damage by anoxia or by lack of glucose. The increased neuronal activity present in the immature brain may also make it more susceptible to seizure activity. It is well known that the seizure threshold is lower in children than in adults. Stimuli that do not cause convulsions in the adult, such as elevation in body temperature, frequently do so in the child. Febrile seizures become less frequent after the age of 5 or 6, that is to say, at about the age when synaptic density begins to decrease in most cortical areas. The advantages of the adult system probably outweigh the likely loss of plasticity related to diffuse representation.

### Plasticity and Complexity of Function

Plasticity appears to be a property of complex neural systems. Very simple nervous systems appear to develop entirely according to fixed genetic programs. In mammalian cerebral cortex, the extent to which cortical functions can be molded by the environment appears to be proportional to the complexity of the anatomic substrate and of its function. Overproduction of synapses during development, a likely index of plasticity, is most marked in primates (both humans and monkeys), and appears to be smaller in cats and in rodents. In the human, environmental effects are more limited in brain regions with relatively simple, basic functions, such as the motor and primary sensory cortex, and they become more important in cortical areas that subserve more complex functions such as speech. Many so-called higher cortical functions do not develop at all without extensive environmental stimulation. Reading is a good example of this. The child learns to read only if taught, and a person may go through life without ever acquiring this skill. The literate person develops a "reading cortex" in the left parietal lobe (the angular gyrus). In the illiterate this cortical region is unable to process written letters and words. We do not know whether the angular gyrus in illiterates subserves different functions, not represented in this region in those with reading skills.

### Crowding Effects

The concept of equipotentiality has one other implication that needs to be mentioned, namely the possibility of overloading or crowding of a system that has to process information related to several different tasks. Competition between the input of the two eyes for cortical representation—a form of crowding—has been observed in binocular interactions in the visual cortex: the representation of one eye in the visual cortex may be increased and that of the opposite eye decreased in response to asymmetric visual input during infancy. Crowding effects are likely to be more marked when less cortical volume is available, as in the case of unilateral brain damage. They have been implicated in subjects with extensive unilateral brain damage and in those with hemispherectomy. Crowding effects may lead to diminished overall functional capacity, as measured by IQ tests. As already noted by Hebb, the major effect of perinatal brain damage tends to be diffuse, manifested by decreased intelligence, rather than loss of specific cortical functions, such as language. This is quite different from the effect of adult brain damage, in which specific cognitive deficits rather than global dementia tend to predominate, and in which IQ is less affected by focal brain lesions. In neonatal brain injury, the cerebral cortex seems to reorganize so as to accommodate most of the basic cortical functions. However, each function appears to be performed somewhat less efficiently. The single hemisphere that is left intact is able to carry out most basic tasks, but is able to do so less well than the bihemispheric brain. While mean IQ is diminished after hemispherectomy, when compared to the normal IQ, this does not preclude the occasional occurrence of high intelligence after hemispherectomy, since variance in IQ after hemispherectomy, as in the normal population, is quite large. Later-developing functions, such as reading, tend to be more impaired than earlier functions such as speech (Banich et al., 1990), which raises the question whether insufficient cortical space is left to accommodate the later function. This view is supported by the finding that IQ decreases during late childhood in subjects with perinatal focal damage to the cerebral cortex (Banich et al., 1990).

The organization of human cerebral cortex differs from that of subhuman species, even primates, in that many functions have their primary representation in only one hemisphere. This unilateral representation is often on the left, as it is for speech functions,[1] although some functions, such as music and prosody (the musical inflection of language), seem to have primarily right-sided representation. This leaves large cortical regions, especially in the right cerebral hemisphere, that are unspecified or silent. These silent areas may become enlisted for a given function when there is increased demand, related either to increased environmental input or to focal brain damage. The existence of these silent areas may give the human brain a unique degree of malleability.

The available evidence, presented in Chapter 6, suggests that lateralization of function in human cerebral cortex develops gradually in the infant and child. For spoken language, unilateral representation in speech areas in the left cerebral hemisphere appears at about the age of 20 months. It occurs later for more complex speech (analysis of sentences) than for simple language tasks (listening to single words). Lateralization of at least some language functions appears to occur earlier than the loss of the ability of the right hemisphere to take over language processing. During a long period after the age of 20 months, extending to late childhood, the speech functions of the right hemisphere may be recovered, as for example when the left hemisphere speech areas are damaged by stroke. We do not know the neural basis of this recovery process. An attractive hypothesis is that circuits for language processing persist in the nondominant hemisphere until late childhood, but do so in an inactive or inhibited state.

A similar delay between "apparent" loss of a function and "irretrievable" loss occurs in speech sound discrimination. Sound discriminations that are not used in the native language appear to be lost between the ages of 6 and 12 months. Sound discriminations peculiar to a given language may be recovered when this language is learned prior to mid-adolescence, but usually

---

1. Recent fMRI data show that predominant right hemisphere activation during a word generation task occurred in only 10 percent of left-handed normal subjects and in no right-handers, which suggests that it occurs in only about 1 percent of normals (Pujol et al., 1999). Bilateral activation was more common in left-handers.

not at a later age. Up to adolescence, a second language is usually learned without accent. That is to say, the sound discriminations used by native speakers are utilized by the second language speakers. The neural basis of this effect probably is similar to that for recovery of language after stroke in the child. It may involve the persistence of neural circuits in an inactive state, and their reactivation by practice. It is known that disuse of a synapse will decrease synaptic strength (the ability of a nerve impulse arriving at the axon terminal to produce depolarization of the postsynaptic dendritic membrane). Conversely, repeated stimulation of a synapse will enhance synaptic transmission (Kandel, Schwartz, and Jessell, 1991). Through such a mechanism, inactive systems may become reactivated when the primary information-processing system is damaged or when there are new functional demands. Some of these inactive systems seem to disappear around early adolescence, perhaps because of the synaptic pruning that occurs at that age. Early adolescence may therefore be an important milestone in brain development, an age at which many early-formed, but unused synaptic circuits are eliminated. Some cortical functions that are used infrequently may be lost in this process. One such function may be childhood memories.

### Childhood Amnesia and Synaptic Pruning

A 5- or 6-year-old has many memories of earlier childhood events and experiences. These are lost by the early adult years, when most individuals have little or no residual memory of early childhood events. Sigmund Freud, a keen observer of child development, described the loss of childhood memories in an essay on "screen memories" (1962, p. 303): "the subject of childhood memories brings into striking relief a fundamental difference between the psychological functioning of children and of adults. No one calls into question the fact that the experiences of the earliest years of our childhood leave ineradicable traces in the depth of our minds. If, however, we seek in our memories to ascertain what were the impressions that were destined to influence us to the end of our lives, the outcome is either nothing at all or a relatively small number of recollections which are often of dubious or enigmatic importance. It is only from the sixth or seventh year onwards, in many cases only after the tenth year, that our lives can be reproduced in memory as a connected chain of events."

Close to 50 percent of all synaptic contacts in the neocortex are lost during late childhood and early adolescence. Synapses that are unspecified for function or those incorporated into circuits that are relatively inactive are most likely to be pruned. These may include circuits involved in the retrieval of stored long-term memories. It would be surprising if the massive pruning that occurs in late childhood were not associated with some loss of cortical function. Childhood memories may well be an example of such loss.

Another likely candidate for functions that are impaired by synaptic pruning is second language retention. While second languages are learned rapidly by the young child, they tend to be forgotten just as fast when not practiced, as has been noted by many immigrants who have taught their native tongue to their children early on, only to have the children forget the second language during the school years. Second language learning by a young child is apt to be of little long-term value unless the language is used frequently after initial acquisition, in the home or in school. This observation contains an important lesson: skills learned early in life, such as mastery of a second language, need to be practiced frequently after they are acquired, or else they are likely to be lost.

### Time Windows for Cortical Plasticity

Recently, the concept of critical periods has been expanded to include periods in brain development during which the effects of environmental stimulation on brain structure and function are maximal. In particular, there has been interest in "time windows of opportunity" (Bateson, 1979), during which teaching and enrichment programs are apt to be most effective. The concept of a time window for learning that shuts suddenly and early, in late infancy, has little scientific support, as was pointed out by Bruer (1999). The time windows that do exist tend to be much wider and tend to close slowly. These windows of opportunity do not coincide with the critical periods for human brain development as defined by Dobbing (1981). Dobbing describes a "vulnerable" period that includes the first 6 or so postnatal months in the human infant. It is characterized by rapid increase in brain weight, which in turn is due largely to the growth of the neuropil (cortical dendritic trees, axonal branches, and synaptic contacts). During this time, the brain is especially susceptible to damage from deprivation of nutrients and of metabolic substrates (glucose and oxygen), from hormonal disturbances, especially deprivation of thyroid hormone, and from exposure to neurotoxic substances such as phenylalanine in phenylketonuria. At this stage of brain development, the effects of environmental stimulation appear to be as yet quite limited. Environmental effects are seen later in development, starting at about the age of 2 months in the human infant in the primary sensory areas, and probably late in the first year or in the second year of life in the association cortex, such as the prefrontal cortex. The period of enhanced effect of environmental input appears to coincide with the plateau of high synaptic density and number that is present from late infancy to late childhood in most cortical areas, except for the primary sensory areas, which show earlier synapse elimination.

The neuroanatomic and neurobiologic findings show no evidence for the existence of an early critical period for learning, extending from fetal life to about the age of 3 years, during which an enriched environment must be provided for optimum brain development. Bruer (1999) has pointed out the lack of evidence for such an early critical period. This does not mean that the early age period is unimportant. Much learning goes on in the early months of life that is likely to provide a basis for future learning and personality development. The point is merely that there is no neurobiologic basis for singling out this age period as having characteristics that make it uniquely competent for certain types of learning. Enhanced stimulation of the fetus, newborn, or young infant has not been shown to affect later competence. Excessive stimulation of the newborn infant might even be detrimental. Before recommending "environmental enrichment" of the fetus or neonate, one would have to carry out carefully designed studies to exclude the negative effects of various types of stimulation, as well as to show that any benefit accrues.

The limits of the time windows of opportunity during which environmental changes *are* likely to have increased effectiveness vary widely, depending on the task. For example, forcing a child to use the squinting eye is most effective in preventing amblyopia (blindness of the squinting eye) when implemented between the ages of 6 months and 4 or 5 years. Second language learning, to be accent free, has to be implemented by early adolescence. Music training has to start prior to the age of 10 years for the musician to be likely to acquire absolute pitch. The differences in these time windows are probably related to differences in the time course of the anatomical development of specific cortical regions.

The evidence, presented in Chapter 2, indicates a hierarchical pattern for synaptic development in the human. Synaptogenesis occurs earliest in the primary visual, auditory, and probably other sensory areas. It occurs later in regions in the parietal and temporal cortex that have to do with sensory integration, language processing, and mathematical and musical competence, and probably latest in the prefrontal cortex, including Broca's motor speech area. Synaptic development appears to depend on the presence of dendritic and axonal growth, which proceeds according to a chronological pattern similar to that seen in synaptogenesis. Myelination of axons traveling to and from the cerebral cortex also is heterochronous. Both the onset of function and the period of increased plasticity can be related to specific stages in the anatomical development of the cerebral cortex in several cortical areas, including the primary visual (calcarine) cortex, Wernicke's speech area, and Broca's area. The approximate time windows for optimum development of several functions are indicated in Table 10.1, p. 212.

Windows of opportunity have recently been widely discussed in relation to the education of children and to the optimum age for introduction of subjects into the curriculum. For example, second (and third) language teaching has been proposed for preschool children. There have been claims that prenatal stimulation, as by playing music to the fetus, is of importance for later learning. What evidence is there in favor of these claims? We know in general that an enriched environment has a positive effect on the child's learning, social adjustment, and even on eventual IQ scores. Animal studies suggest that enrichment programs exert their effect through the formation of new synaptic connections, and that the capacity for new synapse formation persists in the adult, at least to some extent. Longer periods of enrichment, covering the preschool and early school-age years, show larger more sustained effects than 1- or 2-year preschool programs. The effects of enrichment programs are especially evident when children from lower socioeconomic backgrounds are studied. There is some evidence for saturation or ceiling effects. In a child who already lives in a stimulating environment, preschool head start programs appear to have limited impact. Large effects of the home environment and of enrichment programs are also seen in children with developmental disabilities, such as children with anoxic brain damage related to premature birth. A poor environment appears to interact with mild brain damage to yield a poor outcome as far as eventual levels of cognitive and language functions are concerned. In this group of children, early environmental stimulation may make the difference between a person who will eventually function in normal society and one that will require a protected living situation.

# 10

## THE PRACTICAL RELEVANCE OF THE FINDINGS FROM
## DEVELOPMENTAL NEUROBIOLOGY

> Without effort he had learned English, French, Portuguese, Latin.
> Nevertheless—he was not very capable of thought. To think is to for-
> get a difference, to generalize, to abstract.
>
> Jorge Luis Borges, 1962

What have we learned from developmental neurobiology that we can apply to education programs for children? When it comes to specific recommendations as to what should be included in the curriculum at any given age, there are some relevant data from neurobiology, but these need to be applied in conjunction with information from other sources. Certain types of training are likely to be more effective during the period of increased synaptic density and number in cortical regions that are relevant to the task. This information could be utilized in the design of school curricula. For example, second language teaching and musical training are likely to be more effective if started early, during the period of high plasticity, which includes the early school years (ages 5–10 years). Evidence from several sources indicates that this is so: during early puberty, when synaptic density in language areas falls to adult levels, perfect second language acquisition becomes difficult at best, and appears to be impossible for many, who learn the second language with a heavy accent and imperfect grammar. Recent fMRI studies suggest that the functional localizations for second languages differ, depending on the age at which the language was learned. A separate region for second language representation is found more commonly if the second language is learned later (after the age of 8 years), while the second lan-

*Table 10.1.* Time windows of opportunity (optimum periods) for various functions.

| Function | Age |
| --- | --- |
| Recovery of facial movements after stroke | Fetus to neonate |
| Reversal of strabismic amblyopia | 1–5 years |
| Acquisition of absolute pitch | Up to age 10 years |
| Recovery of language after stroke | Up to age 8 years |
| Accent-free second language learning | Up to early adolescence |

guage shares the speech areas of the primary language when it is acquired during the early childhood years. In the more mature brain, the ability to incorporate the second language into the primary language areas appears to be impaired. A new "second language cortex" is formed near the primary speech areas. This area appears to be less efficient than the primary one for language processing. Similarly, the available evidence indicates that the learning of music performance and of music composition is easier when teaching is started early in life, prior to the age of 8–10 years. Some special skills important for musicians, such as development of absolute pitch (Chapter 7), may not be learnable after that age (see Table 10.1).

The early teaching of subjects that are not practiced later is likely to be of limited value. Knowledge acquired early in childhood that is not practiced is apt to be forgotten, especially during late childhood, when many earlier memories appear to be erased ("childhood amnesia"). The neurobiology of development teaches an important lesson: synaptic circuits become less effective and eventually disappear unless utilized. This process may be accelerated in late childhood and early adolescence, when many of the early-formed synaptic contacts in the neocortex disappear. This points to the importance of continued review and use of the early-acquired knowledge base, such as for instance a second language or arithmetic skills. Teaching a second language during grammar school, without provision for subsequent exposure to the language, is apt to be a useless exercise. Yet such early, isolated language training, without provision for later practice and use, has recently been incorporated in some school programs.

Parents and children often object to the many repetitions that appear to be built into most school programs. However, the importance of frequent recapitulation and of continued practice receives support from knowledge of developmental neurobiology. The abandonment of practicing would be apt to increase the loss of previously acquired information, especially during late childhood and early adolescence. At the same time, such review should not be rote repetition. We know from computer simulation of learning in a neural net that the most effective input is varied, but has internal consistency.

Specific applications of neuroscience research to human development are beginning to emerge from the study of input-related plasticity. The new brain imaging techniques are becoming so sensitive that subtle effects of training on the functional organization of the cerebral cortex can be assessed. It therefore is becoming possible to monitor the effects of early intervention programs on information processing by the immature brain, and to identify programs with specific effects on the functional organization of the cerebral cortex. Similarly, the negative effects of early deprivation on brain development can be monitored. The same is true for drug effects. We presently give stimulant medications to thousands of children with so-called attention deficit–hyperactivity disorders, some as young as 3 years old, with little knowledge of the long-term effects—positive or negative— of these medications on the organization of cortical function. The same is true for other psychoactive drugs, including antidepressants and anti-psychotic medications. Functional imaging such as fMRI makes it possible to monitor the effects of these agents on brain development much more closely.

Neuroscience findings also are beginning to be applied to the rehabilitation of children with focal brain injuries. New methods of rehabilitation of infants with brain or peripheral nerve damage, based on neurobiologic principles, are being explored. A good example is the rehabilitation of patients with hemiparesis due to unilateral damage to the central motor pathways. The concept of competition for synaptic sites has led to programs that involve intermittent restraint of the normal hand, and thereby reinforce practice by the paretic hand. In the future, one can foresee the exploration of methods for modification of time windows for anatomical reorganization after focal brain lesions. One such approach involves the modification of the age limit for regrowth of damaged central axons, by inhibition of the control system that normally shuts down the growth of central axons (Chapter 1). Similarly, as we learn more about the control of neurogenesis and synaptogenesis, there is the possibility of influencing the formation and stabilization of neurons and synapses in infants, children, or even adults with neuronal loss secondary to brain trauma or disease. Recent successes with fetal brain cell transplants indicate the feasibility of such approaches. The discovery that transplanted fetal neurons become incorporated into existing neural systems by forming connections with such systems is especially exciting (Piccini et al., 2000). Memory mechanisms, related to acquisition, storage, and retrieval, are another area of intense study that may lead to approaches to the treatment of impaired memory, one of the most common neuro-cognitive symptoms.

An important point remains to be made, namely that neither the biological nor the behavioral sciences provide information as to what *should* be taught to a child. The fact that foreign language acquisition is most complete prior to early adolescence does not mean that one or more foreign languages should be taught to every child. The same point could be made about music teaching. The nature of the school curriculum needs to be determined—at least in part—by other considerations, including the type of knowledge that is likely to promote a richly varied, successful, and happy adult existence, and the availability and prioritization of resources.

One also has to keep in mind that much is as yet unknown concerning the effects of variations in input on brain development. A new "enrichment" program should not be accepted uncritically, without careful study of its effects, both positive and potentially negative. For example, one has to consider the possibility that very ambitious early enrichment and teaching programs may lead to crowding effects and to an early decrease in the size and number of brain regions that are largely unspecified and that may be necessary for creativity in the adolescent and adult. The neurobiologic studies of crowding effects introduce the caveat that too much early learning may under some conditions become detrimental to the learning of later acquired skills. It may be no accident that Albert Einstein was a rather average student in his early years. An extreme example of superior, very early development of some skills and deficiency in others is found in so-called idiot savants. These children may develop astounding computational abilities ("lightning calculators"), but may function in the mentally retarded range on other tasks. The cause of this condition is unknown. However, our present knowledge of developmental neurobiology would suggest the hypothesis that, for some as yet unknown reason, the idiot savant develops functional specification for the preferred task in large brain areas that normally are utilized for other functions.

The brain of the young child may need some "time out" to consolidate information. Reservation of cortical space for the processing of later-acquired skills may also impart a functional advantage. In this regard, it is of interest that U.S. school systems have often been criticized for having longer vacations and for teaching less early on than many European and Asian systems. Yet the United States is a world leader in the production of large numbers of creative young adults in all areas of endeavor. A proper balance of early exposure to an enriched academic environment and time off may be important for optimum cortical development. At this point, we do not know where this balance lies.

In the past decade, there has been a virtual explosion of knowledge in human developmental neurobiology, largely driven by the development of functional imaging techniques, including fMRI, PET, and magnetoencephalography. No one can deny the importance of this work or its relevance to our understanding of infant and child development. At the same time, there has been some concern recently that data from developmental neurobiology have to date provided little scientific evidence in favor of specific recommendations related to child-rearing practices, school curricula, and enrichment programs. Such concern seems unwarranted. As pointed out above, some tentative recommendations can be made, including those based on biologically determined critical periods or windows of opportunity, as long as one keeps in mind that these critical periods are not absolute and that the windows of opportunity do not shut suddenly or tightly. Perhaps "optimum periods" would be a preferable term, since it would allow for the persistence of certain types of learning beyond the "optimum period,'" albeit with increased effort and with imperfect results. However, data derived from developmental neuroscience should never be the sole deciding factor. A balanced view is likely to emerge from the synthesis of results from neuroscience, the behavioral sciences, and educational research.

We are presently at the beginning of an exciting period in developmental neurobiology that will continue for many decades into the twenty-first century, and that is likely to have considerable impact on our approach to education of the normal child as well as the child with neurodevelopmental handicaps.

# REFERENCES

Abrahams, S., A. S. Pickering, C. E. Polkey, and R. G. Morris. 1997. "Spatial memory deficits in patients with unilateral damage to the right hippocampal formation." *Neuropsychologia, 35* (1): 11–24.

Adelson, P. D., D. A. Hovda, J. R. Villablanca, and K. Tatsukawa. 1995. "Development of a crossed corticotectal pathway following cerebral hemispherectomy in cats: a quantitative study of the projecting neurons." *Brain Res., Dev. Brain Res.,* 86: 81–93.

Aghajanian, G., and F. E. Bloom. 1967. "The formation of synaptic junctions in developing rat brain: a quantitative electron microscopic study." *Brain Res.,* 6: 716–727.

Aglioti, S., F. Cortese, and C. Franchini. 1994. "Rapid sensory remapping in the adult human brain as inferred from phantom breast perception." *Neuroreport,* 5: 473–476.

Alho, K., T. Kujala, P. Paavilainen, H. Summala, and R. Naatanen. 1993. "Auditory processing in visual brain areas of the early blind: evidence from event-related potentials." *Electroencephalogr. Clin. Neurophysiol.,* 86: 418–427.

Allendoerfer, K. L., and C. J. Shatz. 1994. "The subplate, a transient neocortical structure: its role in the development of connections between thalamus and cortex." *Ann. Rev. Neurosci.,* 1: 185–218.

Andermann, F. 1991. *Chronic Encephalitis and Epilepsy: Rasmussen Syndrome.* Stoneham, Mass.: Butterworth-Heinemann.

Anderson, B., and T. Harvey. 1996. "Alterations in cortical thickness and neuronal density in the frontal cortex of Albert Einstein." *Neurosci. Lett.,* 210: 161–164.

Anglin, J. M. 1993. "Vocabulary development: a morphological analysis." *Monogr. Soc. Res. Child Dev.,* 58 (10): 1–166.

Aram, D. M., B. Ekelman, and H. Whitaker. 1985. "Verbal and cognitive sequelae following unilateral lesions acquired in early childhood." *J. Clin. Exp. Neuropsychol.,* 7: 55–78.

——— 1986. "Spoken syntax in children with acquired unilateral hemisphere lesions." *Brain Lang.,* 27: 75–100.

——— 1987. "Lexical retrieval in left and right brain-lesioned children." *Brain Lang.,* 28: 61–87.

Armand, J., E. Olivier, S. A. Edgley, and R. N. Lemon. 1997. "Postnatal development of corticospinal projections from motor cortex to the cervical enlargement in the Macaque monkey." *J. Neurosci.,* 17: 251–266.

Armstrong, D., J. K. Dunn, B. Antalffi, and R. Trivetti. 1995. "Selective dendritic alterations in the cortex of Rett Syndrome." *J. Neuropath. Exp. Neurol.,* 54: 195–201.

Armstrong-James, M., and R. Johnson. 1970. "Quantitative studies of postnatal changes in synapses in rat superficial motor cerebral cortex: an electron microscopical study." *Z. Zellforsch Mikrosk Anat.,* 11: 559–568.

Arnold, G. L., R. S. Kirby, S. Langendoerfer, and L. Wilkins-Haug. 1994. "Toluene embryopathy: clinical delineation and developmental follow-up." *Pediatrics,* 93: 216–220.

Asanuma, C., and B. B. Stanfield. 1990. "Induction of somatic sensory inputs to the lateral geniculate nucleus in congenitally blind mice and in phenotypically normal mice." *Neuroscience,* 39:533–545.

Asbury, A. K., G. M. McKhann, and W. I. McDonald. 1992. *Diseases of the Nervous System.* 2nd ed. Philadelphia: W. B. Saunders.

Assaf, A. A. 1982. "The sensitive period: transfer of fixation after occlusion for strabismic amblyopia." *Br. J. Opthalmol.,* 66: 64–70.

Bach-y-Rita, P. 1981. "Brain plasticity as a basis of the development of rehabilitation procedures for hemiplegia." *Scand. J. Rehabil. Med.,* 13: 73–83.

Banich, M. T., S. C. Levine, H. Kim, and P. Huttenlocher. 1990. "The effects of developmental factors on IQ in hemiplegic children." *Neuropsychologia,* 28: 35–47.

Barkovich, A. J. 1990. "Normal development of the neonatal and infant brain." In *Pediatric Neuroimaging*, ed. A. J. Barkovich, 5–34. New York: Raven Press.

Barnea, A., and F. Nottebohm. 1994. "Seasonal recruitment of hippocampal neurons in adult free-ranging Black-Capped Chickadees." *Proc. Natl. Acad. Sci. USA*, 91: 11217–11221.

Basser, L. S. 1962. "Hemiplegia of early onset and the faculty of speech, with special reference to the effects of hemispherectomy." *Brain*, 85: 427–460.

Bates, C. H., and H. P. Killackey. 1984." The emergence of a discretely distributed pattern of corticospinal projection neurons." *Brain Res.*, 375: 265–273.

Bates, E. 1993. "Comprehension and production in early language development." *Monogr. Soc. Res. Child Dev.*, 58: 222–242.

Bateson, P. 1979. "How do sensitive periods arise and what are they for?" *Animal Behavior*, 27: 470–486.

Baumann, M. L., and T. L. Kemper. 1982. "Morphologic and histoanatomic observations of the brain in untreated human phenylketonuria." *Acta Neuropathol.*, 58: 55–63.

Baynes, K. 1990. "Language and reading in the right hemisphere: highways and byways of the brain?" *J. Cogn. Neurosci.*, 2: 159–179.

Beard, J. 1896. On Certain Problems of Vertebrate Embryology." Jena: Gustav Fisher.

Beaulieu, C., H. D'Arceuil, M. Hedehus, A. de Crespigny, A. Kastrup, and M. E. Mosley. 1999. "Diffusion-weighted magnetic resonance imaging: theory and potential applications to child neurology." *Seminars in Pediatr. Neurol.*, 6: 87–100.

Becker, L. E., D. L. Armstrong, F. Chan, and M. M. Wood. 1984. "Dendritic development in human occipital cortical neurons." *Dev. Brain Res.*, 13: 117–124.

Behrman, R. E., and V. C. Vaughan III, ed. 1987. *Nelson's Textbook of Pediatrics*. 13th ed. Philadelphia: W. B. Saunders.

Benson, R. R., W. J. Logan, G. R. Cosgrove, A. J. Cole, H. Jiang, L. L. LeSueur, B. R. Buchbinder, B. R. Rosen, and V. S. Caviness, Jr. 1996. "Functional MRI localization of language in a 9-year-old child." *Can. J. Neurol., Sci.* 23: 213–219.

Berman, N. E. J. 1991. "Alterations in visual cortical connections in cats following early removal of retinal input." *Dev. Brain Res.*, 63: 163–180.

Berry, R. J., Z. Li, J. D. Erickson, S. Li, C. A. Moore, H. Wang, J. Mulinare, P. Zhao, L. Y Wong, J. Gindler, S. X. Hong, and A. Correa. 1999. "Prevention of neural tube defects with folic acid in China." *N. Engl. J. Med.*, 341: 1485–1490.

Best, C. T. 1988. "The emergence of cerebral asymmetries in early human development: a literature review and a neuroembryological model." In *Brain Lateralization in Children*, ed. D. L. Molfese and S. J. Segalowitz, 5–34. New York: The Guilford Press.

Bhide, P. G., and D. O. Frost. 1992. "Axon substitution in the reorganization of developing neural connections." *Proc. Natl. Acad. Sci. USA*, 89: 11847–11851.

Binder, J. R., J. A. Frost, T. A. Hammeke, R. W. Cox, S. M. Rao, and T. Prieto. 1997. "Human brain language areas identified by functional magnetic resonance imaging (fMRI)." *Neuroscience*, 17: 353–362.

Bishop, D. V. M. 1983. "Linguistic impairment after left hemidecortication for infantile hemiplegia: a reappraisal." *Q. J. Exp. Psychol.*, 35A: 199–207.

Black, P., R. S. Markowitz, and S. N. Cianci. 1975. "Recovery of motor function after lesions in motor cortex of monkeys." In *Outcome of Severe Damage to the Central Nervous System*, ed. P. Black, 65–83. New York: Elsevier.

Blakemore, C. 1977. "Genetic instructions and developmental plasticity in the kitten's visual cortex." *Phil. Trans. Roy. Soc. Lond. B.*, 278: 425–434.

Blakemore, C., and R. C. Van Sluyters. 1975. "Innate and environmental factors in the development of the kitten's visual cortex." *J. Physiol.* (London), 248: 663–716.

Bloom, F. E., and G. K. Aghajanian. 1968. "Fine structural and cytochemical analysis of the staining of synaptic junctions with phosphotungstic acid." *J. Ultrastruct. Res.*, 22: 361–375.

Bonn, D. 1998. "Tune in early for best results with cochlear implants." *The Lancet*, 352: 1836.

Borges, J. L. 1962. "Funes, the memorious." In *Ficciones*. New York: Grove Press.

Bourgeois, J. P., P. S. Goldman-Rakic, and P. Rakic. 1994. "Synaptogenesis in the prefrontal cortex of Rhesus monkeys." *Cereb. Cortex*, 4: 78–96.

Bourgeois, J. P., P. Jastreboff, and P. Rakic. 1989. "Synaptogenesis in the visual cortex of normal and preterm monkeys: evidence for intrinsic regulation of synaptic overproduction." *Proc. Natl. Acad. Sci. USA*, 86: 4297–4301.

Bourgeois, J. P., and P. Rakic. 1993. "Changes of synaptic density in the primary visual cortex of the Macaque monkey from fetal to adult stage." *J. Neurosci.*, 13: 2801–2820.

——— 1996. "Synaptogenesis of the occipital cortex in Macaque monkey devoid of retinal input from early embryonic stages." *Eur. J. Neurosci.*, 8: 942–950.

Braak, H., and E. Braak. 1985. "Golgi preparations as a tool in neuropathology with particular reference to investigations of the human telencephalic cortex." *Prog. Neurobiol.*, 25: 93–139.

Brodal, A. 1973. "Self-observation and neuro-anatomical considerations after a stroke." *Brain*, 96: 675–694.

Brody, B. A, H. C. Kinney, A. S. Kloman, and F. H. Gilles. 1987. "Sequence of central nervous system myelination in human infancy. I. An autopsy study of myelination." *J. Neuropathol. Exp. Neurol.*, 46: 283–301.

Brooks-Gunn, J., R. T. Gross, H. C. Kraemer, D. Spiker, and S. Shapiro. 1992. "Enhancing the cognitive outcomes of low birth weight premature infants: for whom is the intervention most effective?" *Pediatrics*, 89: 1209–1215.

Brooks-Gunn, J., P. K. Klebanov, F. Liaw, and D. Spiker. 1993. "Enhancing the development of low birthweight, premature infants: changes in cognition and behavior over the first three years." *Child Dev.*, 64: 736–753.

Brown, M. C., J. K. S. Jansen, and D. Van Essen. 1976. "Polyneuronal innervation of skeletal muscles in newborn rats and its elimination during maturation." *J. Physiol.*, 261: 387–422.

Bruer, J. T. 1999. *The Myth of the First Three Years*. New York: Free Press.

Buell, S. J., and P. D. Coleman. 1981. "Quantitative evidence for selective dendritic growth in normal aging, but not in senile dementia." *Brain Res.*, 214: 23–31.

Buisseret, P., and P. Imbert. 1976. "Visual cortical cells: their developmental properties in normal and dark-reared kittens." *J. Physiol.* (London), 255: 511–525.

Buonomano, D. V., and M. M. Merzenich. 1998. "Cortical plasticity: from synapses to maps." *Ann. Rev. Neurosci.*, 21: 149–186.

Burgess, J. W., and J. R. Villablanca. 1986. "Recovery of function after neonatal or adult hemispherectomy in cats. II. Limb bias and development; paw usage, locomotion, and rehabilitative effects of exercise." *Behav. Brain Res.*, 20: 1–18.

Burgess, J. W., J. R. Villablanca, and M. S. Levine. 1986. "Recovery of function after neonatal or adult hemispherectomy in cats. Complex functions: activity, aggression, sociality, maze, and holeboard performance." *Behav. Brain Res.*, 20: 217–230.

Calford, M. B., and R. Tweedale. 1990. "Interhemispheric transfer of plasticity in the cerebral cortex." *Science*, 249: 805–807.

Campbell, F. A., and C. T. Ramey. 1994. "Effects of early intervention on intellectual and academic achievement: a follow-up study of children from low-income families." *Child Dev.*, 65: 684–698.

Campbell, F. A., and C. T. Ramey. 1995. "Cognitive and school outcomes for high-risk African-American students at middle adolescence: positive effects of early intervention." *Am. Edu. Res. J.*, 32: 742–772.

Cao, Y., V. L. Towle, D. N. Levin, and J. M. Balter. 1993. "Functional mapping of human motor cortical activation with conventional MR imaging at 1.5T." *J. Magn. Reson. Imaging*, 3:869–875.

Cao, Y., E. M. Vikingstad, P. R. Huttenlocher, V. L. Towle, S. C. Levine, and D. N. Levin. 1994. "Functional magnetic resonance studies of the reorganization of the hand sensorimotor area after unilateral brain injury in the perinatal period." *Proc. Natl. Acad. Sci. USA*, 91: 9612–9616.

Carlson-Kuhta, P., J. R. Villablanca, and L. D. Loopuijt. 1997. "Innervation of the caudate nucleus, thalamus, and red nucleus by the remaining sensorimotor cortex in cats with fetal or neonatal unilateral frontal cortex removal." *Brain Res., Dev. Brain Res.*, 98: 234–246.

Casey, B. J., R. J. Trainor, J. L. Orendi, A. B. Schubert, L. E. Nystrom, J. N. Giedd,

F. X. Castellanos, J. V. Haxby, D. C. Noll, J. D. Cohen, S. D. Forman, R. E. Dahl, and J. L. Rapoport. 1997. "A developmental functional MRI study of prefrontal activation during performance of a go–no-go task." *J. Cogn. Neurosci.*, 9: 835–847.

Castro, A. J. 1975. "Ipsilateral corticospinal projections after large lesions of the cerebral hemisphere in neonatal rats." *Exp. Neurol.*, 46: 1–8.

Changeux, J. P. 1980. "Genetic determinism and epigenesis of the neural network: is there a biological compromise between Chomsky and Piaget?" In *Language and Learning: The Debate between Jean Piaget and Noam Chomsky*, ed. M. Piattelli-Palmarini, 185–197. Cambridge, Mass.: Harvard University Press.

Changeux, J. P., and A. Danchin. 1976. "Selective stabilization of developing synapses as a mechanism for the specification of neuronal networks." *Nature*, 264: 705–712.

Changeux, J. P., T. Heidmann, and P. Patte. 1984. "Learning by selection." In *The Biology of Learning*, ed. P. Marler and H. S. Terrace, 115–137. New York: Springer-Verlag.

Chen, R., C. Gerloff, J. Classen, E. Wassermann, M. Hallett, and L. G. Cohen. 1997. "Safety of different inter-train intervals for repetitive transcranial magnetic stimulation and recommendations for safe ranges of stimulation parameters." *Electroencephalogr. Clin. Neurophysiol.*, 105: 415–421.

Chi, F. G., E. C. Dooling, and F. H. Gilles. 1977. "Left-right asymmetries of the temporal speech areas of the human fetus." *Arch. Neurol.*, 34: 346–348.

Chu, D., P. R. Huttenlocher, D. N. Levin, and V. L. Towle. 2000. "Reorganization of the hand somatosensory cortex following perinatal unilateral brain injury." *Neuropediatrics*, 31: 63–69.

Chugani, D. C., O. Muzik, M. Behen, R. Rothermel, J. J. Janisse, J. Lee, and H. T. Chugani. 1999. "Developmental changes in serotonin synthesis capacity in autistic and non-autistic children." *Ann. Neurol.*, 45: 287–295.

Chugani, H. T., M. E. Behen, O. Muzik, C. Juhasz, F. Nagy, and D. C. Chugani. 2001. "Local brain functional activity following early deprivation: a study of postinstitutionalized Romanian orphans." *Neuroimage*, 14: 1290–1301.

Chugani, H. T., E. da Silva, and D. C. Chugani. 1997. "PET in the diagnostic evaluation of children with focal epilepsy." In *Pediatric Epilepsy Syndromes and Their Surgical Treatment,* ed. I. Tuxhorn, H. Holtthausen, and H. Boenigk, 592–606. London: John Libbey.

Chugani, H. T., and M. E. Phelps. 1986. "Maturational changes in cerebral function in infants determined by 18FDG positron emission tomography." *Science,* 231: 840–843.

——— 1990. "Imaging human brain development with positron emission tomography." *J. Nucl. Med.,* 32: 23–25.

Chugani, H. T., M. E. Phelps, and J. C. Mazziotta. 1987. "Positron emission tomography study of human brain functional development." *Ann. Neurol.,* 22: 487–497.

Chun, J. J., and C. J. Shatz. 1989. "Interstitial cells of the adult neocortical white matter are the remnant of the early-generated subplate neuron population." *J. Comp. Neurol.,* 282: 555–569.

Clark, G. D., M. Mizuguchi, B. Antalffy, J. Barnes, and D. Armstrong. 1997. "Predominant localization of the LIS family of gene products to Cajal-Retzius cells and ventricular neuroepithelium in the developing human cerebral cortex." *J. Neuropath. Exp. Neurol.,* 56: 1044–1052.

Cohen, L. G., P. Celnik, A. Pascual-Leone, B. Corwell, L. Faiz, J. Dambrosia, M. Honda, N. Sedato, C. Gerloff, M. D. Catala, and M. Hallett. 1997. "Functional relevance of cross-modal plasticity in blind humans." *Nature,* 389: 180–183.

Collen, F. M., D. T. Wade, and C. M. Bradshaw. 1990. "Mobility after stroke: reliability of measures of impairment and disability." *Int. Disabil. Studies,* 12: 6–9.

Conel, J. 1939–1963. *The Postnatal Development of the Human Cerebral Cortex.* 6 vols. Cambridge, Mass.: Harvard University Press.

Cooper, N. G. F., and D. A. Steindler. 1986. "Lectins demarcate the barrel sub-field in the somatosensory cortex of the early postnatal mouse." *J. Comp. Neurol.,* 249: 157–169.

Courchesne, E. 1978. "Neurophysiological correlates of cognitive development: changes in long-latency event-related potentials from childhood to adulthood." *Electroencephalogr. Clin. Neurophysiol.,* 45: 468–482.

Cowan, W. M., J. W. Fawcett, D. D. O'Leary, and B. B. Stanfield. 1984. "Regressive events in neurogenesis." *Science,* 225: 1258–1265.

Cowan, W. M., and E. Wenger. 1967. "Cell loss in the trochlear nucleus of the chick during normal development and after radical extirpation of the optic vesicle." *J. Exp. Zool.,* 164: 265–280.

Cragg, B. G. 1975a. "The density of synapses and neurons in normal, mentally defective, and aging human brains." *Brain,* 98: 81–90.

———— 1975b. "The development of synapses in the visual system of the cat." *J. Comp. Neurol.,* 160: 265–280.

———— 1975c. "The development of synapses in kitten visual cortex during visual deprivation." *Exp. Neurol.,* 46: 445–451.

Crain, S. 1991. "Language acquisition in the absence of experience." *Behav. Brain Sci.,* 14: 597–611.

Cravioto, J., and E. R. Delicardie. 1970. "Nutrition, growth, and development. Mental performance in school age children. Findings after recovery from early severe malnutrition." *Amer. J. Dis. Child.,* 120: 404–410.

Crawford, M. L., J. T. de Faber, R. S. Harwerth, E. L. Smith III, and G. K. von Noorden. 1989. "The effects of reverse monocular deprivation in monkeys. II. Electrophysiological and anatomical studies." *Exp. Brain Res.,* 74: 338–347.

Crome, L., and J. Stern. 1972. *Pathology of Mental Retardation.* 2nd ed. Edinburgh and London: Churchill Livingstone.

Curtiss, S. 1977. *Genie: A Psycholinguistic Study of a Modern-Day Wild Child.* New York: Academic Press.

Czeizel, A. E., and I. Dudas. 1992. "Prevention of the first occurrence of neural tube defects by periconceptional vitamin supplementation." *N. Engl. J. Med.,* 327: 1832–1835.

D'Amato, C., and S. P. Hicks. 1978. "Normal development and post-traumatic plasticity of corticospinal neurons in rats." *Exp. Neurol.,* 60: 557–569.

Darian-Smith, C., and C. D. Gilbert. 1994. "Axonal sprouting accompanies functional reorganization in adult cat striate cortex". *Nature,* 368: 737–740.

Davison, A. N., and J. Dobbing. 1968. "The developing brain." In *Applied Neurochemistry,* ed. A. N. Davison and J. Dobbing, 253–263. Oxford: Blackwell.

Daw, N. 1995. *Visual Development*. New York: Plenum Press.

———— 1997. "Critical periods and strabismus: what questions remain?" *Optometry and Vision Science,* 74: 690–694.

Dehaene, S., E. Spelke, P. Pinel, R. Stanescu, and S. Tsivkin. "Sources of mathematical thinking: behavioral and brain imaging evidence." *Science,* 284: 970–974.

Dehay, C., J. Bullier, and H. Kennedy. 1984. "Transient projections from the fronto-parietal and temporal cortex to areas 17, 18, and 19 in the kitten." *Exper. Brain Res.,* 57: 208–212.

Dehay, C., G. Horsburgh, M. Berland, H. Killackey, and H. Kennedy. "Maturation and connectivity of the visual cortex in monkey is altered by prenatal removal of retinal input." *Nature,* 337: 265–267.

Dennis, M., and B. Kohn. 1975. "Comprehension of syntax in infantile hemiplegias after cerebral hemidecortication: left hemisphere superiority." *Brain Lang.,* 2: 472–482.

Dennis, M., and H. Whitaker. 1976. "Language acquisition following hemidecortication: linguistic superiority of the left over the right hemisphere." *Brain Lang.,* 3: 404–433.

De Pedro-Cuesta, J., L. Widen-Holmqvist, and P. Bach-y-Rita. 1992. "Evaluation of stroke rehabilitation by randomized controlled studies: a review." *Acta Neurol. Scand.,* 86:433–439.

Derbyshire, S. W., B. A. Vogt, and A. K. Jones. 1998. "Pain and Stroop interference tasks activate separate processing modules in anterior cingulate cortex." *Exp. Brain Res.,* 118: 52–60.

De Volder, A. J., A. Bol, J. Blin, A. Robert, P. Arno, C. Grandin, C. Michel, and C. Veraart. 1997. "Brain energy metabolism in early blind subjects: neural activity in the visual cortex." *Brain Res.,* 750: 235–244.

Diamond, A. 1985. "The development of the ability to use recall to guide action, as indicated by infants performance on AB." *Child Dev.,* 56: 868–883.

Diamond, A., and P. S. Goldman-Rakic. 1989. "Comparison of human infants and Rhesus monkeys on Piaget's AB task: evidence for dependence on dorso-lateral prefrontal cortex." *Exp. Brain Res.,* 74: 24–40.

Diamond, M. C., F. Law, H. Rhodes, B. Lindner, M. R. Rosenzweig, D. Krech, and

E. L. Bennett. 1966. "Increases in cortical depth and glia numbers in rats subjected to enriched environment". *J. Comp. Neurol.*, 128: 117–126.

Diamond, M. C., A. B. Scheibel, J. G. M. Murphy, and T. Harvey. 1985. "On the brain of a scientist: Albert Einstein." *Exp. Neurol.*, 88: 198–204.

Diebler, M. F., E. Farkas-Bargeton, and R. Wehrle. 1979. "Developmental changes of enzymes associated with energy metabolism and the synthesis of some neurotransmitters in discrete areas of human neocortex." *J. Neurochem.*, 32: 429–435.

Dobbing, J. 1981. "The later development of the brain and its vulnerability." In *Scientific Foundations of Pediatrics*, 2nd ed., ed. J. A. Davis and J. Dobbing, 744–759. London: William Heinemann Medical Books.

Dobbing, J., and J. Sands. 1970. "Timing of neuroblast multiplication in developing human brain." *Nature*, 226: 639–640.

———— 1973. "Quantitative growth and development of human brain." *Arch. Dis. Child.*, 48: 757–767.

———— 1979. "Comparative aspects of the brain growth spurt." *Early Hum. Dev.*, 3: 79–83.

Dodd, J., and T. M. Jessell. 1988. "Axon guidance and the patterning of neuronal projections in vertebrates." *Science*, 242: 692–699.

Doetsch, F, J. M. Garcia-Verdugo, and A. Alvarez-Buylla. 1997. "Cellular composition and three-dimensional organization of the subventricular germinal zone in the adult mammalian brain." *J. Neurosci.*, 17 (13): 5046–5061

Doughtery, D. D., A. A. Bonab, T. J. Spencer, S. L. Rauch, B. K. Madras, and A. J. Fischman. 1999. "Dopamine transporter density with attention deficit hyperactivity disorder." *The Lancet*, 354: 2132–2133.

Doupe, A. J. 1996. "Plasticity of a different feather?" *Science*, 274: 1851–1853.

———— 1997. "Song and order-selective neurons in the songbird anterior forebrain and their emergence during vocal development." *J. Neurosci.*, 17: 1147–1167.

Doupe, A. J., and P. K. Kuhl. 1999. "Birdsong and human speech: common themes and mechanisms." *Ann. Rev. Neurosci.*, 22: 567–631.

Doupe, A. J., and M. M. Solis. 1997. "Song- and order-selective neurons develop in the songbird anterior forebrain during vocal learning." *J. Neurobiol.,* 33: 694–709.

Eayrs, J. T. 1971. "Thyroid and developing brain: anatomical and behavioral effects." In *Hormones in Development,* ed. M. Hamburg and E. J. Barrington, 345–355. New York: Appleton-Century-Crofts.

Edelman, G. M. 1987. *Neural Darwinism: The Theory of Neuronal Group Selection.* New York: Basic Books.

Eisele, J. and D. Aram. 1994. "Comprehension and imitation of syntax following early hemisphere damage." *Brain Lang.,* 46: 212–231.

——— 1995. "Lexical and grammatical development in children with early hemisphere damage: a cross-sectional view from birth to adolescence." In *The Handbook of Child Language,* ed. P. Fletcher and B. MacWhinney, 664–689. Oxford: Basil Blackwell.

Elbert, T., C. Pantev, C. Wienbruch, B. Rockstroh, and E. Taub. 1995. "Increased cortical representation of the fingers of the left hand in string players." *Science,* 270: 305–307.

Elman, J. L. 1993. "Learning and development in neural networks: the importance of starting small." *Cognition,* 48: 71–99.

Elman, J. L., E. A. Bates, M. H. Johnson, A. Karmiloff-Smith, D. Parisi, and K. Plunkett. 1996. *Rethinking Innateness: A Connectionist Perspective on Development.* Cambridge, Mass.: MIT Press.

Elman, J. L., and D. Zipser. 1988. "Learning the hidden structure of speech." *J. Acoust. Soc. Am.,* 83: 1615–1626.

Eriksson, P., E. Perfilieva, T. Bjork-Eriksson, A. M. Albom, C. Nordborg, D. A. Peterson, and F. H. Gage. 1998. "Neurogenesis in the adult human hippocampus." *Nature Medicine,* 4: 1313–1317.

Evarts, E. V., Y. Shinoda, and S. P. Wise. 1984. *Neurophysiological Approach to Higher Brain Function.* New York: J. Wiley.

Fawcett, J. W., and D. D. M. O'Leary. 1985. "The role of electrical activity in the formation of topographical maps." *Trends in Neurosci.,* 8: 201–206.

Fedrick, J., and A. B. M. Anderson. 1976. "Factors associated with spontaneous preterm birth." *Brit. J. Obstet. Gynecol.,* 83: 342–350.

Feinberg, I., H. C. Thode, Jr., H. T. Chugani, and J. D. March. 1990. "Gamma distribution model describes maturational curves for delta wave amplitude, cortical metabolic rate, and synaptic density." *J. Theoretical Biol., 142:* 149–161.

Feldman, D. E., R. A. Nicoll, and R. C. Malenka. 1999. "Synaptic plasticity at thalamocortical synapses in developing rat somatosensory cortex: LTP, LTD, and silent synapses." *J. Neurobiol.,* 41:92–101.

Fenson, L., P. S. Dale, J. S. Reznick, E. Bates, D. J. Thal, and S. J. Pethick. 1994. "Variability in early communicative development." *Monogr. Soc. Res. Child Dev., 59 (5):* 1–173.

Fisher, R. S., R. L. Sutton, D. A. Hovda, and J. R. Villablanca. 1988. "Corticorubral connections: ultrastructural evidence for homotypical synaptic reinnervation after developmental deafferentation." *J. Neurosci. Res.,* 21: 438–446.

Flege, J. E., M. J. Munro, and I. R. A. MacKay. 1995. "Effects of age of second-language learning on the production of English consonants." *Speech Communication,* 16: 1–26.

Flexner, L. B. 1951–52. "The development of the cerebral cortex: a cytological, functional, and biochemical approach." *Harvey Lectures,* 47: 156–179.

Floeter, M. K., and W. T. Greenough. 1979. "Cerebellar plasticity: modification of Purkinje cell structure by differential rearing in monkeys." *Science,* 206: 227–229.

Fox, J. W., E. D. Lamperti, Y. Z. Eksioglu, S. E. Hong, Y. Feng, D. A. Graham, I. E. Scheffer, W. B. Dobyns, B. A. Hirsch, R. Radtje, S. F. Berkovic, P. R. Huttenlocher, and C. A. Walsh. 1998. "Mutations in Filamin I prevent migration of cerebral cortical neurons in human periventricular heterotopia." *Neuron,* 21: 1315–1325.

Fredrick, J., and A. B. M. Anderson. 1976. "Factors associated with spontaneous preterm birth." *Brit. J. Obstet. Gynecol.,* 83: 342–350.

French, N. P., R. Hagan, S. F. Evans, M. Godfrey, and J. Newnham. 1999. "Repeated antenatal corticosteroids: size at birth and subsequent development." *Am. J. Obstet. Gynecol.* 180: 114–121.

Freud, S. 1897. *Die Infantile Cerebrallähmung.* Vienna: A. Hölder.

——— 1962. "Screen memories." In *The Standard Edition of the Complete Psychological Works of Sigmund Freud,* vol. 3, 303–322. London: The Hogarth Press.

Friedman, J. M., and J. E. Polifka. 1994. *Teratogenic Effects of Drugs: A Resource for Clinicians.* Baltimore: The Johns Hopkins University Press.

Frost, D. O. 1982. "Anomalous visual connections to somatosensory and auditory systems following brain lesions in early life." *Brain Res.,* 255: 627–635.

——— 1984. "Axonal growth and target selection during development: retinal projections in the ventrobasal complex and other "non-visual" structures in neonatal Syrian hamsters." *J. Comp. Neurol.,* 230: 576–592.

——— 1990. "Sensory processing by novel experimentally induced cross-modal circuits." *Ann. N.Y. Acad. Sci.,* 608: 92–109, discussion on 109–112.

Fryauf-Bertschy, H., R. S. Tyler, D. M. Kelsay, B. J. Gantz, and G. G. Woodworth. 1997. "Cochlear implant use by prelingually deafened children: the influences of age at implant and length of device use." *J. Speech Lang. Hear. Res.,* 40: 183–199.

Galaburda, A. M. 1984. "Anatomical asymmetries." In *Cerebral Dominance: The Biological Foundations,* ed. N. Geschwind and A. M. Galaburda, 11–25. Cambridge, Mass.: Harvard University Press.

Garber, B., P. R. Huttenlocher, and L. H .M. Larramendi. 1980. "Self-assembly of cortical plate cells in vitro within mouse cerebral aggregates." *Brain Res.,* 201: 255–278.

Gauthier, I., A. W. Anderson, M. J. Tarr, P. Skudlarski, and J. C. Gore. 1997. "Levels of categorization in visual recognition studied using functional magnetic resonance imaging." *Current Biology,* 7: 645–651.

Gauthier, I. and M. J. Tarr. 1997. "Becoming a 'Greeble' expert: exploring mechanisms for face recognition." *Vision Research,* 37: 1673–1682.

Gauthier, I., P. Williams, M. J. Tarr, and J. Tanaka. 1998. "Training 'greeble' experts: a framework for studying expert object recognition processes." *Vision Research,* 38: 2401–2428.

Geschwind, N., and W. Levitsky. 1968. "Human brain: left/right asymmetries in temporal speech region." *Science,* 161: 186–189.

Gesell, A. 1929. *Infancy and Human Growth.* New York: Macmillan.

Ghosh, A., A. Antonini, S. K. McConnell, and C. J. Shatz. 1990. "Requirement for subplate neurons in the formation of thalamocortical connections." *Nature,* 347: 179–181.

Giedd, J. N., J. M. Rumsey, F. X. Castellanos, J. C. Rajapakse, D. Kaysen, A. C. Vaituzis, Y. C. Vauss, S. D. Hamburger, and J. L. Rapoport. 1996. "A quantitative MRI study of the corpus callosum in children and adolescents." *Dev. Brain Res.,* 91: 274–280.

Goldfield, B. A., and J. S. Reznick. 1990. "Early lexical acquisition: rate, content, and the vocabulary spurt." *J. of Child Lang.,* 17: 171–183.

Goldman, P. S. 1976. "The role of experience in recovery of function following orbital prefrontal lesions in infant monkey." *Neuropsychologia,* 14: 401–412.

——— 1977. "Salutary effects of early experience on deficits caused by lesions of frontal association cortex in developing Rhesus monkeys." *Exp. Neurol.,* 57: 588–602.

——— 1978. "Neuronal plasticity in primate telencephalon: anomalous projections induced by prenatal removal of frontal cortex." *Science,* 202: 768–770.

Goldman, P. S., and T. W. Galkin. 1978. "Prenatal removal of frontal association cortex in the fetal rhesus monkey: anatomical and functional consequences in postnatal life. *Brain Res.,* 152: 451–485.

Golgi, E. 1903. *Opera omnia.* Milan: U. Hoepli.

Gómez-Pinilla, F., J. R. Villablanca, B. J. Sonnier, and M. S. Levine. 1986. "Reorganization of pericruciate cortical projections to the spinal cord and dorsal column nuclei after neonatal or adult cerebral hemispherectomy in cats." *Brain Res.,* 385: 343–355.

Goodman, C. S., and C. J. Shatz. 1993. "Developmental mechanisms that generate precise patterns of neuronal connectivity." *Neuron,* 10: 77–98.

Gospe, S. M., Jr. and S. S. Zhou. 1998. "Toluene abuse embryopathy: longitudinal developmental effects of prenatal exposure to toluene in rats." Reprod. Toxicol., 12: 119–126.

Gould, E., A. Beylin, P. Tanapat, A. Reeves, and T. J. Shors. 1999. "Learning enhances adult neurogenesis in the hippocampal formation." *Nat. Neurosci.,* 2: 260–265.

Gould, E., A. J. Reeves, M. Fallah, P. Tanapat, C. G. Gross, and E. Fuchs. 1999. "Hippocampal neurogenesis in adult Old World primates." *Proc. Natl. Acad. Sci. USA,* 96: 5263–5267.

Gould, E., A. J. Reeves, M. S. A. Graziano, and C. G. Gross. 1999. "Neurogenesis in the neocortex of adult primates." *Science,* 286: 548–552.

Gould, E., and P. Tanapat. 1997. "Lesion-induced proliferation of neural progenitors in the dentate gyrus of the adult rat." *Neuroscience,* 80: 427–436.

Gould, H. J., C. G. Cusick, T. P. Pons, and J. H. Kaas. 1986. "The relationship of corpus callosum connections to electrical stimulation maps of motor, supplementary motor, and the frontal eye fields in owl monkeys." *J. Comp. Neurol.,* 247: 297–325.

Green, E. J., W. T. Greenough, and B. E. Schlumpf. 1983. "Effects of complex or isolated environments on cortical dendrites of middle-aged rats." *Brain Res.,* 264: 233–240.

Greenfield, J. G. 1958. *Neuropathology.* London: Edward Arnold.

Greenough, W. T., J. E. Black, and C. S. Wallace. 1987. "Experience and brain development." *Child Dev.,* 58: 539–559.

Greenough, W. T., and F. L. Chang. 1988. "Dendritic pattern formation involves both oriented regression and oriented growth in the barrels of mouse somatosensory cortex." *Brain Res.,* 471: 148–152.

Greenough, W. T., H. M. F. Hwang, and C. Gorman. 1985. "Evidence for active synapse formation or altered postsynaptic metabolism in visual cortex of rats reared in complex environments." *Proc. Natl. Acad. Sci. USA,* 82: 4549–4552.

Greenough, W. T., and F. R. Volkmar. 1973. "Patterns of dendritic branching in occipital cortex of rats reared in complex environments." *Exp. Neurol.,* 40: 491–504.

Gressens P., B. E. Kosofsky, and P. Evrard. 1992. "Cocaine-induced disturbances of corticogenesis in the developing murine brain." *Neurosci. Lett.,* 140: 113–116.

Gressens P, M. Lammens, J. J. Picard, and P. Evrard. 1992. "Ethanol-induced disturbances of gliogenesis and neurogenesis in the developing murine brain:

an in vitro and in vivo immunohistochemical and ultrastructural study." *Alcohol and Alcoholism,* 3: 219–226.

Grigorenko, E. L., F. B. Wood, M. S. Meyer, L. A. Hart, W. C. Speed, A. Shuster, and D. L. Pauls. 1997. "Susceptibility loci for distinct components of developmental dyslexia on chromosomes 6 and 15." *Amer. J. Hum. Genet.,* 60: 27–39.

Gumbinas M., M. Oda, and P. R. Huttenlocher. 1973. "The effects of corticosteroids on myelination of the developing rat brain." *Biol. Neonate,* 22: 355–366.

Gunderson, C. H., and G. B. Solitare. 1968. "Mirror movements in patients with Klippel-Feil syndrome." *Arch. Neurol.,* 18: 675.

Haier, R. J., B. V. Siegel, Jr., A. Maclachlan, E. Soderling, S. Lottenberg, and M. S. Buchsbaum. 1992. "Regional glucose metabolic changes after learning a complex visuospatial/motor task: a positron emission tomographic study. *Brain Res.,* 570:134–143.

Hamburger, V. 1975. "Cell death in the development of the lateral motor column of the chick embryo. *J. Comp. Neurol,.* 160: 535–546.

Harbord, M. G., J. P. Finn, M. A. Hall-Craggs, S. A. Robb, B. E. Kendall, and S. G. Boyd. 1990. "Myelination patterns on magnetic resonance of children with developmental delay." *Dev. Med. Child Neurol.,* 32: 295–303.

Harding, G. F. A., J. Grose, A. Wilton, and J. G. Bissenden. 1988. "The pattern reversal VEP in short gestation infants." *Electroencephalogr. Clin. Neurophysiol.,* 74: 76–80.

Harris, K. Y. 2000. "Handedness and footedness in Korean college students." *Brain & Cognition,* 43 (1–3): 268–274.

Harwerth, R. S., E. L. Smith III, M. L. Crawford, and G. K. von Noorden. 1989. "The effects of reverse monocular deprivation in monkeys. I. Psychophysical experiments." *Exp. Brain Res.,* 74: 327–347.

Haydon, P. G., and P. Drapeau. 1995. "From contact to connection: early events during synaptogenesis." *Trends in Neurosci.,* 18: 196–210.

Hebb, D. O. 1949. *The Organization of Behavior.* New York: J. Wiley.

Hein, A., and R. Held. 1967. "Dissociation of the visual placing response into elicited and guided components." *Science,* 158: 390–391.

Heiss, W. D., J. Kessler, A. Thiel, M. Ghaemi, and H. Karbe. 1999. "Differential capacity of left and right hemispheric areas for compensation of post stroke aphasia." *Ann. Neurol.,* 45: 430–438.

Held, R., and J. A. Bauer, Jr. 1967. "Visually guided reaching in infant monkeys after restricted rearing." *Science,* 155: 718–720.

Held, R., E. Nirch, and J. Gwiazda. 1980. "Stereoacuity of human infants." *Proc. Natl. Acad. Sci. USA,* 77: 5572–5574.

Herb, E., and U. Thyen. 1992. "Mutism after cerebellar medulloblastoma surgery." *Neuropediatrics,* 23: 144–146.

Hernnstein, R. J., and C. Murray. 1994. *The Bell Curve: Intelligence and Class Structure in American Life.* New York: Free Press.

Hertz-Pannier, L., W. D. Gaillard, S. Mott, C. A. Cuenod, S. Y. Bookheimer, S. Weinstein, J. Conry, P. H. Papero, S. J. Schiff, D. Le Bihan, and W. H. Theodore. 1997. "Non-invasive assessment of language dominance in children and adolescents with functional MRI." *Neurology,* 48: 1003–1012.

Hessler, N. A., and A. J. Doupe. 1999. "Social context modulates singing-related neural activity in the songbird forebrain." *Nat. Neurosci.,* 2: 209–211.

Heumann, D., and G. Leuba. 1983. "Neuronal death in the development and aging of the cerebral cortex of the mouse." *Neuropath. Appl. Neurobiol.,* 9: 297–311.

Hicks, S. P., and C. J. D'Amato. 1970. "Motor-sensory and visual behavior after hemispherectomy in newborn and mature rats." *Exp. Neurol.,* 29: 416–438.

Hines, T. 1998. "Further on Einstein's brain." *Exp. Neurol.,* 150: 343–344.

Hirata, Y., S. Kuriki, and C. Pantev. 1999. "Musicians with absolute pitch show distinct neural activities in the auditory cortex. *Neuroreport,* 6: 999–1002.

Hirotsune, S., M. W. Fleck, M. J. Gambello, G. J. Bix, A. Chen, G. D. Clark, D. H. Ledbetter, C. J. McBain, and A. Wynshaw-Boris. 1998. "Graded reduction of Pafahl1b1 (Lis1) activity results in neuronal migration defects and early embryonic lethality." *Nature Genetics,* 19: 333–339.

Hitchcock, P. F., and J. L. Hickey. 1980. "Ocular dominance columns: evidence for their presence in humans." *Brain Res.*, 182: 176–179.

Hobbelen, J. F., and M. W. Van Hof. 1986. "Short communication: the hopping reaction in the rabbit after early and late removal of the motor cortex." *Behav. Brain Res.*, 21: 73–77.

Hoff-Ginsberg, E. 1998. "The relation of birth order and socioeconomic status to children's language experience and language development." *Applied Psycholinguistics*, 19: 603–629.

Holden, C. 1998. "Mozart for Georgia newborns". *Science*, 279: 663.

Hollyday, M., and V. Hamburger. 1976. "Reduction of the naturally occurring motor neuron loss by enlargement of the periphery." *J. Comp. Neurol.*, 170: 311–320.

Honegger, P., and D. Lenoir. 1980. "Triodothyronine enhancement of neuronal differentiation in aggregating fetal rat brain cells cultured in a chemically defined medium." *Brain Res.*, 199: 425–434.

Hoo Nah, S., L. S. Ong, and S. K. Leong. 1980. "Is sprouting the result of a persistent neonatal connection?" *Neurosci. Lett.*, 19: 39–44.

Horton, J. C., and E. T. Hedley-White. 1984. "Mapping of cytochrome oxidase patches and ocular dominance columns in human visual cortex." *Phil. Trans. Roy. Soc. Lond. B.*, 304: 255–272.

Horton, J. C., and D. R. Hocking. 1996. "An adult-like pattern of ocular dominance columns in striate cortex of newborn monkeys prior to visual experience." *J. Neurosci.*, 15: 1791–1807.

Hovda, D. A., and J. R. Villablanca. 1989. "Depth perception in cats after cerebral hemispherectomy: comparisons between neonatal- and adult-lesioned animals." *Behav. Brain Res.*, 32: 231–240.

——— 1990. "Sparing of visual field perception in neonatal but not adult cerebral hemispherectomized cats: relationship with oxidative metabolism of the superior colliculus." *Behav. Brain Res.*, 37: 119–132.

Hovda, D. A., J. R. Villablanca, H. T. Chugani, and M. E. Phelps. 1996. "Cerebral metabolism following neonatal or adult hemineodecortication in cats: effects on glucose metabolism using [14C] 2-deoxy-D-glucose autoradiography." *J. Cereb. Blood Flow Metab.*, 16: 134–146.

Hubel, D. H., and T. N. Wiesel. 1963a. "Shape and arrangement of columns in the cat's striate cortex." *J. Neurophysiol.*, 165: 559–568.

——— 1963b. "Receptive fields of cells in striate cortex of very young, visually inexperienced kittens." *J. Neurophysiol.*, 26: 944–1002.

——— 1965. "Binocular interactions in striate cortex of kittens reared with artificial squint." *J. Neurophysiol.*, 28: 1041–1059.

——— 1970. "The period of susceptibility to the physiological effects of unilateral eye closure in kittens." *J. Physiol.*, 206: 419–436.

Hubel, D. H., T. N. Wiesel, and S. LeVay. 1977. "Plasticity of ocular dominance columns in the monkey striate cortex." *Phil. Trans.Roy. Soc. Lond*, 278: 377–409.

Hubel, D. H., T. N. Wiesel, and M. P. Stryker. 1978. "Anatomical demonstration of ocular dominance columns in Macaque monkey." *J. Comp. Neurol.*, 177: 361–380.

Hughes, A. F. 1961. "Cell degeneration in the larval ventral horn of Xenopus laevis (Daudin)." *J. Embryol. Exper. Morphol.* 9: 269–284.

Huppi, P. S., S. E. Maier, S. Peled, G. P. Zientara, P. D. Barnes, F. A. Jolesz, and J. J. Volpe. 1998. "Microstructural development of human newborn cerebral white matter assessed in vivo by diffusion tensor magnetic resonance imaging." *Pediatric Research*, 44 (4): 584–590.

Hurst, J. A., M. Baraitser, E. Auger, F. Graham, and S. Norell. "An extended family with a dominantly inherited speech disorder." *Dev. Med. Child Neurol.*, 32: 347–355.

Huttenlocher, J., W. Haight, A. Bryk, M. Seltzer, and T. Lyons. 1991. "Early vocabulary growth: relation to language input and gender." *Dev. Psych.*, 27: 236–248.

Huttenlocher, J., S. Levine, and J. Vevea. 1998. "Environmental input and cognitive growth: a study using time-period comparisons." *Child Dev.*, 69: 1012–1029.

Huttenlocher, P. R. 1970. "Myelination and the development of function in immature pyramidal tract." *Exp. Neurol.*, 29: 405–415.

——— 1974. "Dendritic development in neocortex of children with mental defect and infantile spasms." *Neurology*, 24: 203–210.

—— 1979. "Synaptic density in human frontal cortex: developmental changes and effects of aging." *Brain Res.,* 163: 195–205.

—— 1990. "Morphometric study of human cerebral cortex development." *Neuropsychologia,* 28: 517–527.

—— 1991. "Dendritic and synaptic pathology in mental retardation." *Pediatr. Neurol.,* 7: 79–85.

—— 1994. "Synaptogenesis, synapse elimination, and neural plasticity in human cerebral cortex." In *Threats to Optimal Development: Integrating Biological, Psychological, and Social Risk Factors: The Minnesota Symposia on Child Psychology,* ed. Charles A. Nelson, vol. 27, 35–54. L. Erlbaum: Mahwah, N.J.

—— 2000. "Synaptogenesis in human cerebral cortex and the concept of critical periods." In *The Role of Early Experience in Infant Development,* ed. N. A. Fox, L. A. Leavitt, and J. G. Warhol, 15–28. St. Louis: Johnson & Johnson Pediatric Institute.

Huttenlocher, P. R., and A. S. Dabholkar. 1997. "Regional differences in synaptogenesis in human cerebral cortex." *J. Comp. Neurol.,* 387: 167–178.

Huttenlocher, P. R., and C. deCourten. 1987. "The development of synapses in striate cortex of man." *Human Neurobiol.,* 6: 1–9.

Huttenlocher, P. R., C. deCourten, L. G. Garey, and H. VanderLoos. 1982. "Synaptogenesis in human visual cortex: evidence for synapse elimination during normal development." *Neurosci. Lett.,* 33: 247–252.

Huttenlocher, P. R., and P. T. Heydemann. 1984. "The fine structure of cortical tubers in tuberous sclerosis." *Ann. Neurol.,* 16: 592–602.

Huttenlocher, P. R., and R. M. Raichelson. 1989. "Effects of neonatal hemispherectomy on location and number of corticospinal neurons in the rat." *Dev. Brain Res.,* 47: 59–69.

Innocenti, G. M. 1981. "Growth and reshaping of axons in the establishment of visual callosal connections." *Science,* 212: 824–827.

—— 1995. "Exuberant development of connections, and its possible permissive role in cortical evolution." *Trends in Neurosci.,* 18: 397–402.

Innocenti, G. M., and S. Clarke. 1984. "Bilateral transitory projection to visual areas from auditory cortex in kittens." *Dev. Brain Res.,* 14: 143–148.

Innocenti, G. M., L. Fiore, and R. Caminiti. 1977. "Exuberant projection into the corpus callosum from the visual cortex of newborn rats." *Neurosci. Lett.,* 4: 237–242.

Ipiña, S. L., A. Ruiz-Marcos, F. Escobar del Rey, and G. M. de Escobar. 1987. "Pyramidal cortical cell morphology studied by multivariate analysis: effects of neonatal thyroidectomy, aging, and thyroxine-substitution therapy." *Dev. Brain Res.,* 37: 219–229.

Isaac, J. T., M. C. Crair, R. A. Nicoll, and R. C. Malenka. 1997. "Silent synapses during development of thalamocortical inputs." *Neuron,* 18: 269–280.

Jackson III, G. F., J. R. Villablanca, and L. D. Loopuijt. 1995. "Few neocortical and thalamic morphological changes after a neonatal frontal cortical ablation contrast with the effects of a similar lesion in fetal cats." *Brain Res., Dev. Brain Res.,* 90: 62–72.

Jacobson, M. 1991. *Developmental Neurobioly.* 3rd ed. New York and London: Plenum Press.

Jancke, L., G. Schlaug, and H. Steinmetz. 1997. "Hand skill asymmetry in professional musicians." *Brain Cogn.,* 34: 424- 432.

Jenny, A. B., and C. B. Saper. 1987. "Organization of the facial nucleus and cortico-facial projection in the monkey: a reconsideration of the upper motor neuron facial palsy. *Neurology,* 37: 930–939.

Jernigan, T. L., D. A. Trauner, J. R. Hesselink, and P. A. Tallal. 1991. "Maturation of human cerebrum observed in vivo during adolescence." *Brain,* 114: 2037–2049.

Johnson, J. S., and M. Newport. 1989. "Critical period effects in second language learning: the influence of maturational state on the acquisition of English as second language." *Cogn. Psychol.,* 21: 60–99.

Johnston, M. V., and J. T. Coyle. 1981. "Development of central neurotransmitter systems." In *The Fetus and Independent Life,* ed. K. Elliot and J. Whelan, 251–270. London: Pittman.

Johnston, M. V., F. S. Silverstein, F. O. Reindel, J. B. Penney, Jr., and A. B. Young. 1985. "Muscarinic cholinergic receptors in human infant forebrain: [$^3$H] quinuclidinyl benzilade binding in homogenates and quantitative autoradiography in sections." *Dev. Brain Res.,* 19: 195–203.

Jones, K., and E. W. Smith. 1973. "Recognition of the fetal alcohol syndrome in early infancy." *The Lancet,* 2: 999–1001.

Joosten, E. A., A. A. Gribnau, and P. J. Dederen. 1987. "An anterograde tracer study of the developing corticospinal tract in the rat: three components." *Dev. Brain Res.,* 36: 121–130.

Julesz, B., and I. Kovacs, ed. 1995. *Maturational Windows and Adult Cortical Plasticity.* Reading, Mass.: Addison Wesley.

Just, M. A., P. A. Carpenter, T. A. Keller, W. F. Eddy, and K. R. Thulborn. 1996. "Brain activation modulated by sentence comprehension." *Science,* 274: 114–116.

Kalil, K., and T. Reh. 1979. "Regrowth of severed axons in the neonatal central nervous system: establishment of normal connections." *Science,* 205: 1158–1161.

———— 1982. "A light and electron microscopic study of regrowing pyramidal tract fibers." *J. Comp. Neurol.,* 211: 265–275.

Kandel, E. R., J. H. Schwartz, and T. M. Jessell. 1991. *Principles of Neural Science.* 3rd ed. New York: Elsevier.

Kanwisher, N., J. McDermott, and M. M. Chun. 1997. "The fusiform face area: a module in human extrastriate cortex specialized for face perception." *J. Neurosci.,* 17: 4302–4311.

Karni, A., G. Meyer, P. Jezzard, M. Adams, R. Turner, and L. Ungerleider. 1995. "Functional MRI evidence for adult motor cortex plasticity during motor skill learning." *Nature,* 377: 155–158.

Kasten, E., S. Wust, W. Behrens-Baumann, and B. A. Sabel. 1998. "Computer-based training for the treatment of partial blindness." *Nature Medicine,* 4: 1083–1087.

Kempermann, G., H. G. Kuhn, and F. H. Gage. 1997." More hippocampal neurons in adult mice living in an enriched environment". *Nature,* 386: 493–495.

Kennard, M. A. 1940. "Relation of age to motor impairment in man and in subhuman primates." *Arch. Neurol. Psychiatr.,* 44: 377–397.

———— 1942. "Cortical reorganization of motor function studies on series of monkeys of various ages from infancy to maturity." *Arch. Neurol. Psychiatr.,* 48: 227–240.

Kennedy, C., and L. Sokoloff. 1957. "An adaptation of the nitrous oxide method to the study of the cerebral circulation in children: normal values for cerebral blood flow and cerebral metabolic rate in childhood." *J. Clin. Invest.,* 36: 1130–1137.

Killackey, H. P. 1990."Neocortical expansion: an attempt toward relating phylogeny and ontogeny." *J. Cogn. Neurosci.,* 2: 1–17.

Kim, K. H. S., N. R. Relkin, K.-M. Lee, and J. Hirsch. 1997. "Distinct cortical areas associated with native and second languages." *Nature,* 388: 171–174.

Kim, Y. H., D. R. Gitelman, A. C. Nobre, T. B. Parrish, K. S. LaBar, and M. M. Mesulam. 1999. "The large-scale neural network for spatial attention displays multifunctional overlap, but differential asymmetry." *Neuroimage,* 9: 269–277.

Kinney, H. C., B. A. Brody, A. S. Kloman, and F. H. Gilles. 1988. "Sequence of central nervous system myelination in human infancy. II. Patterns of myelination in autopsied infants." *J. Neuropathol. Exp. Neurol.,* 47: 217–234.

Kleim, J. H., E. Lussnig, E. R. Schwartz, T. A. Comery, and W. T. Greenough. 1996. Synaptogenesis and Fos expression in the motor cortex of the adult rat after motor skill learning. *J. Neurosci.,* 16: 4929–4935.

Klekamp, J., A. Riedel, C. Harper, and H. J. Kretschmann. 1991. "Quantitative changes during the postnatal maturation of the human visual cortex." *J. Neurosci.,* 103: 136–143.

Klingberg, T., C. J. Vaidya, J. D. Gabrieli, M. E. Moseley, and M. Hedehus. 1999. "Myelination and organization of the frontal white matter in children: a diffusion tensor MRI study." *Neuroreport,* 10 (13): 2817–2821.

Koch, S., G. Losche, E. Jager-Roman, S. Jacob, D. Raing, A. Deichl, and H. Helge. 1992. "Major and minor birth malformations and antiepileptic drugs." *Neurology,* 42 (suppl. 5): 83–88.

Koenderink, M. J., and H. B. Uylings. 1996. "Morphometric dendritic field analysis of pyramidal neurons in the human prefrontal cortex: relation to section thickness." *J. Neurosci. Methods,* 64: 115–122.

Kohn, B. and M. Dennis. 1974. "Selective impairments of visuospatial abilities in infantile hemiplegics after right cerebral hemidecortication." *Neuropsychologia,* 12: 505–512.

Kolb, B., and I. Q. Whishaw. 1989. "Plasticity in the neocortex: mechanisms under-lying recovery from early brain damage." *Prog. Neurobiol.* 32: 235–276.

——— 1998. "Brain plasticity and behavior." *Ann. Rev. Psych.*, 49:, 43–64.

Konishi, M. 1995. "A sensitive period for birdsong learning." In *Maturational Windows and Adult Cortical Plasticity*, ed. B. Julesz and I. Kovacs, 87–92. Santa Fe Institute Studies in the Sciences of Complexity. Reading, Mass.: Addison Wesley.

Kornhuber, J., F. Mack-Burghardt, C. Konradi, J. Fritze, and P. Riederer. 1989. "Effects of antemortem and postmortem factors on MK 801 binding in the human brain." *Life Sciences*, 45: 745–749.

Kornhuber, J., W. Retz, P. Riederer, H. Heinsen, and J. Fritze. 1988. "Effects of antemortem and postmortem factors on glutamate binding in the human brain." *Neurosci. Lett.*, 93: 312–317.

Kostovic, I., N. Lukinovic, M. Judas, N. Bogdanovic, L. Mrzljak, N. Zecevic, and M. Kubat. 1989. "Structural basis of the developmental plasticity in the human cerebral cortex: the role of the transient subplate zone." *Metab. Brain Dis.*, 4: 17–23.

Kostovic I., and M. E. Molliver. 1974. "A new interpretation of the laminar development of cerebral cortex: synaptogenesis in different layers of neopallium in the human fetus." *Anat. Rec.*, 178: 395.

Kostovic, I., and P. Rakic. 1990. "Developmental history of transient subplate zone in the visual and somatosensory cortex of the Macaque monkey and human brain." *J. Comp. Neurol.*, 297: 441–470.

Kostovic, I., L. Seress, L. Mrzljak, and M. Judas. 1989. "Early onset of synapse formation in the human hippocampus: a correlation with Nissl-Golgi architectonics in 15- and 16.5-week-old fetuses." *Neuroscience*, 30: 105–116.

Kuffler, S. W. 1953. "Discharge patterns and functional organization of mammalian retina." *J. Clin. Neuropsychol.*, 16: 37–68.

Kuhl, P. K. 1994. "Learning and representation in speech and language." *Current Opinion in Neurobiol.*, 4: 812–822.

Kuhl, P. K., J. E. Andruski, I. A. Chistovich, L. A. Chistovich, E. V. Kozhevnikova, V. L. Ryskina, E. I. Stolyarova, U. Sundberg, and F. Lacerda. 1997. "Cross-language analysis of phonetic units in language addressed to infants." *Science*, 277: 686–686.

Kuhl, P. K., K. A. Williams, F. Lacerda, K. N. Stevens, and B. Lindblom. 1992. "Linguistic experience alters phonetic perception in infants by 6 months of age." *Science,* 255: 606–608.

Kuida K, T. F., C. Y. Haydar, Y. Kuan, Y. Gu, C. Taya, H. Karasuyama, M. S. Su, P. Rakic, and R. A. Flavell. 1998. "Reduced apoptosis and cytochrome c-mediated caspase activation in mice lacking caspase 9." *Cell,* 95: 325–327.

Kujala, T., K. Alho, P. Paavilainen, H. Summala, and R. Naatanen. 1990. "Neural plasticity in processing of sound location by the early blind: an event-related potential study." *Electroencephalogr. Clin. Neurophysiol.,* 84: 469–472.

Kwakkel, G., R. C. Wagenaar, J. W .R. Twisk, G. J. Lankhorst, and J. C. Koetsier. 1999. "Intensity of leg and arm training after primary middle-cerebral-artery stroke: a randomized trial." *The Lancet,* 354: 191–196.

Lafuente, M. J., R. Grifol, J. Segarra, and J. Soriano. 1996. "Effects of the Firstart method of prenatal stimulation on psychomotor development: the first six months." *Pre- & Peri-Natal Psychology J.,* 11: 151–162.

LaMantia, A. C., and P. Rakic. 1990. "Axon overproduction and elimination in the corpus callosum of the developing Rhesus monkey." *J. Neurosci.,* 10: 2156–2175.

Lampe, J. B., S. M. Touch, and A. R. Spitzer. 1999. "Repeated antenatal steroids: size at birth and subsequent development." *Clin. Pediatr.,* 38, 553–554.

Landry, P., and M. Deschênes. 1981. "The extensive arborization of incoming afferents to cortex." *J. Comp. Neurol.,* 199: 345.

Lane, H. 1976. *The Wild Boy of Aveyron.* Cambridge, Mass.: Harvard University Press.

Larramendi, L. M. H. 1969. "Analysis of synaptogenesis in the cerebellum of the mouse." In *Neurobiology of Cerebellar Evolution and Development,* ed. R. Llinas, 803–843. Chicago: AMA Education and Research Foundation.

Lashley, K. S. 1929. *Brain Mechanisms and Intelligence: A Quantitative Study of Injuries to the Brain.* Chicago: University of Chicago Press.

——— "In search of the engram." In *Physiological Mechanisms and Animal Behavior.* Symposia of the Society for Experimental Biology, no. 4. New York: Academic Press.

——— 1951. *Central Mechanisms in Behavior.* New York: J. Wiley.

Lehmann, J. F., B. J. DeLateur, R. S. Fowler, Jr., C. G. Warren, R. Arnhold, G. Schertzer, R. Hurka, J. J. Whitmore, A. J. Masock, and K. H. Chambers. 1975. "Stroke: does rehabilitation affect outcome?" *Arch. Phys. Rehabil.,* 56: 375–382.

Leng, X., and G. L. Shaw. 1991. "Toward a neural theory of higher brain function using music as a window." *Concepts in Neurosci.,* 2: 229–258.

Lenn, N. J., and A. J. Freinkel. 1989. "Facial sparing as a feature of prenatal-onset hemiparesis." *Pediatr. Neurol.,* 5: 291–295.

Lenneberg, E. A. 1967. *Biological Foundations of Language.* New York: J. Wiley.

Leong, S. K., and R. Lund. 1973. "Anomalous bilateral corticofugal pathways in albino rats after neonatal lesions." *Brain Res.,* 62: 218–221.

Leuba, G., and L. J. Garey. 1987. "Evolution of neuronal numerical density in the developing and aging human visual cortex." *Human Neurobiol.,* 6: 11–18.

Leuba, G., and R. Kraftsik. 1994. "Changes in volume, surface estimate, three-dimensional shape, and total number of neurons of the human primary visual cortex from midgestation until old age." *Anat. Embryol.,* 190: 351–366.

LeVay, S., M. P. Stryker, and C. J. Shatz. 1978. "Ocular dominance columns and their development in layer IV of the cat's visual cortex: a quantitative study." *J. Comp. Neurol.,* 179: 223–244.

LeVay, S., T. N. Wiesel, and D. H. Hubel. 1980. "The development of ocular dominance columns in normal and visually deprived monkeys." *J. Comp. Neurol.,* 19: 11–51.

Leventhal, A. G., and H. V. B. Hirsch. 1975. "Cortical effects of early selective exposure to vertical lines." *Science,* 190: 102–104.

Levine, S. C., P. Huttenlocher, M. T. Banich, and E. Duda. 1987. "Factors affecting cognitive functioning of hemiplegic children." *Dev. Med. Child Neurol.,* 29: 27–35.

Logan, W. J. 1999. "Functional magnetic resonance imaging in children." *Seminars in Pediatr. Neurol.,* 6: 78–86.

Loopuijt, L. D., D. A. Hovda, A. Ebrahim, J. R. Villablanca, and H. T. Chugani. 1998. "Differences in D2 dopamine receptor binding in the neostriatum between cats hemidecorticated neonatally or in adulthood." *Brain Res., Dev. Brain Res.,* 107, 113–122.

Lorente de No, R. 1938. "Cerebral cortex: architecture, intracortical connections, and motor projections." In *Physiology of the Nervous System,* ed. J. F. Fulton, 288–331. New York: Oxford University Press.

MacArthur, B. A., R. N. Howie, J. A. Dezoete, and D. A. Elliott. 1982. "School progress and cognitive development of 6-year-old children whose mothers were treated antenatally with betamethasone." *Pediatrics,* 70: 99–105.

Maguire, E., N. Burgess, and J. O'Keefe. 1999. "Human spatial navigation: cognitive maps, sexual dimorphism, and neural substrates." *Current Opinion in Neurobiol.,* 9: 171–177.

Maguire E. A., D. G. Gadian, I. S. Johnsrude, C. D. Good, J. Ashburner, R. S. J. Frackowiak, and C. D. Frith. 2000. "Navigation-related structural change in the hippocampi of cab drivers." *Proc. Natl. Acad. Sci., USA,* 97: 4398–4403.

Marin-Padilla, M. 1970. "Prenatal and early postnatal ontogenesis of the human motor cortex: a Golgi study. I. The sequential development of the cortical layers." *Brain Res.,* 23: 167–183.

Martins, I. P., and J. M. Ferro. 1992. "Recovery of acquired aphasia in children." *Aphasiology,* 6: 431–438.

Maurer, D., and T. L. Lewis. 1993. "Visual outcomes after infantile cataract." In *Early Visual Development, Normal and Abnormal,* ed. K. Simons, 454–484. New York: Oxford University Press.

Maurer, D., T. L. Lewis, H. P. Brent, and A. V. Levin. 1999. "Rapid improvement in the acuity of infants after visual input." *Science,* 286: 108–110.

Maw, R., J. Wilks, I. Harvey, T. J. Peters, and J. Golding. 1999. "Early surgery compared with watchful waiting for glue ear and effect on language development in preschool children: a randomized trial." *The Lancet,* 353: 960–963.

Merzenich, M. M., R. J. Nelson, M. P. Stryker, M. S. Cynader, A. Schoppmann, and J. M. Zook. 1984. "Somatosensory cortical map changes following digit amputation in adult monkeys". *J. Comp. Neurol.,* 224: 591–605.

Mesulam, M. M. 1990. "Large-scale neurocognitive networks and distributed processing for attention, language ,and memory." *Ann. Neurol.,* 28: 597–613.

Michel, A. E., and L. J. Garey. 1984. "The development of dendritic spines in the human visual cortex." *Human Neurobiol.,* 3: 223–227.

Miller, M. W. 1986. "Effects of alcohol on the generation and migration of cerebral cortical neurons." *Science,* 233: 1308–1311.

Miller, M. W., and G. Potempa. 1990. "Numbers of neurons and glia in a mature rat somatosensory cortex: effects of prenatal exposure to ethanol." *J. Comp. Neurol.,* 293: 92–102.

Miller, M. W., and S. Robertson. 1993. "Prenatal exposure to ethanol alters the postnatal development and transformation of radial glia to astrocytes in the cortex." *J. Comp. Neurol.,* 337: 253–266.

Mills, D. L., S. Coffey-Corina, and H. J. Neville. 1993. "Language acquisition and cerebral specialization in 20-month-old infants." *J. Cogn. Neurosci.,* 5: 317–334.

Mills, D. L., S. Coffey-Corina, and H. J. Neville. 1997. "Language comprehension and cerebral specialization from 13–20 months." *Dev. Psychol.,* 13: 397–445.

Ming, J. E., and M. Muenke. 1998. "Holoprosencephaly: from Homer to hedgehog." *Clini. Genetics,* 53: 155–163.

Mohindra, I., S. G. Jacobson, and R. Held. 1983. "Binocular visual form deprivation in human infants." *Documenta Ophthalmologica,* 55: 237–249.

Molfese, D., and J. Betz. 1988. "Electrophysiological indices of the early development of lateralization of language and cognition, and their implication for predicting later development." In *Brain Lateralization in Children: Developmental Implications,* ed. D. L. Molfese and S. J. Segalowitz, 171–190. New York: Guilford.

Molliver, M. E., I. Kostovic, and H. VanderLoos. 1973. "The development of synapses in the human fetus." *Brain Res.,* 50: 403–407.

Mountcastle, V. B., and I. Darian-Smith. 1968. "Neural mechanisms in somesthesia." In *Medical Physiology,* 12th ed., ed. V. B. Mountcastle, vol. 2., 1372–1423. St. Louis: Mosby.

MRC Vitamin Study Research Group. 1991. "Prevention of neural tube defects: results of the Medical Research Council Vitamin Study." *The Lancet,* 338: 131–137.

Mrzljak, L., H. B. M. Uylings, C. G. Van Eden, and M. Judas. 1990. "Neuronal development in human prefrontal cortex in prenatal and postnatal stages." *Prog. Brain Res.,* 85: 185–222.

Muller, R. A., M. E. Behen, R. D. Rothermel, O. Muzik, P. K. Chakraborty, and H. T. Chugani. 1999. "Brain organization for language in children, adolescents, and adults with left hemisphere lesion: a PET study." *Prog. Neuropsychopharmacol. Biol. Psych.,* 23: 657–668.

Muller, R. A., H. T. Chugani, O. Muzik, and T. J. Mangner. 1998. "Brain organization of motor and language functions following hemispherectomy: a [(15)O]-water positron emission study." *J. Child Neurol.,* 13: 16–22.

Muller, R. A., R. D. Rothermel, M. E. Behen, O. Muzik, P. K. Chakraborty, and H. T. Chugani. 1999. "Language organization in patients with early and late left-hemisphere lesions: a PET study." *Neuropsychologia,* 37: 545–557.

Muller, R. A., R. D. Rothermel, M. E. Behen, O. Muzik, T. J. Mangner, P. K. Chakraborty, and H. T. Chugani. 1998. "Brain organization of language after early unilateral lesion: a PET study." *Brain Lang.,* 62: 422–451.

Muller, R. A., R. D. Rothermel, M. E. Behen, O. Muzik, T. J. Mangner, and H. T. Chugani. 1997. "Receptive and expressive language activations for sentences: a PET study." *Neuroreport,* 8: 3767–3770.

Nantais, K. M., and E. G. Schellenberg. 1999. "The Mozart effect: an artifact of preference." *Psychological Science,* 10: 370–373.

Nass, R. 1985. "Mirror movement asymmetries in congenital hemiparesis: the inhibition hypothesis revisited." *Neurology,* 35: 1059–1062.

Nelson, C. A., and F. E. Bloom. 1997. "Child development and neuroscience." *Child Dev.,* 68: 970–987.

Neville H. J., D. Bavelier, D. Corina, J. Rauschecker, A. Karni, A. Lalwani, A. Brown, V. Clark, P. Jezzard, and R. Turner. 1998. "Cerebral organization for language in deaf and hearing subjects: biological constraints and effects of experience." *Proc. Natl. Acad. Sci. USA,* 95: 922–929.

Neville, H. J., M. Kutas, and A. Schmidt. 1982. "Event-related potential studies of cerebral specification during reading: II. Studies of congenitally deaf adults." *Brain Lang.,* 16: 316–337.

Neville, H. J., and D. Lawson. 1987. "Attention to central and peripheral visual space in a movement detection task. III. Separate effects of auditory deprivation and acquisition of a visual language." *Brain Res.*, 405: 284–294.

Neville, H. J., D. L. Mills, and D. S. Lawson. 1992. "Fractioning language: different neural subsystems with different sensitive periods." *Cereb. Cortex*, 2: 244–258.

Neville, H. J., A. Schmidt, and M. Kutas. 1983. "Altered visual-evoked potentials in congenitally deaf adults." *Brain Res.*, 266: 127–132.

Newport, E. L. 1990. "Maturational constraints on language learning." *Cogn. Science*, 14: 11–28.

Nezu, A., S. Kimura, S. Uehara, T. Kobayashi, M. Tanaka, and K. Saito. 1997. "Magnetic stimulation of motor cortex in children: maturity of corticospinal pathway and problem of clinical application." *Brain Dev.*, 19: 176–180.

Nicholson, J. L., and J. Altman. 1972. "Synaptogenesis in the rat cerebellum: effects of early hypo- and hyperthyroidism." *Science*, 176: 530–532.

Nomura, Y., H. Sakuma, K. Taketa, T. Tugami, Y. Okuda, and T. Nakagawa. 1994. "Diffusional anisotropy of the human brain assessed with diffusion-weighted MR: relation with brain development and aging." *Amer. J. Neuroradiol.*, 15: 231–238.

Nottebohm, F. 1985. "Neuronal replacement in adulthood". *Ann. N.Y. Acad. Sci.*, 457: 143–161.

Ojemann, G. A. 1991. "Cortical organization of language." *J. Neurosci.*, 11: 2281–2287.

O'Kusky, J., and M. Colonnier. 1982. "Postnatal changes in the number of neurons and synapses in the visual cortex A17 of the Macaque monkey." *J. Comp. Neurol.*, 210: 291–296.

O'Leary, D. D. M. 1989. "Do cortical areas emerge from a protocortex?" *Trends in Neurosci.*, 12: 400–406.

——— 1992. "Development of connectional diversity and specificity in the mammalian brain by pruning of collateral projections." *Current Opinion in Neurobiol.*, 2: 70–77.

O'Leary, D. D. M., D. Crespo, J. W. Fawcett, and W. M. Cowan. 1986. "The effect of intraocular tetrodotoxin on the postnatal reduction in numbers of optic nerve axons in the rat." *Dev. Brain Res.*, 30: 96–103.

O'Leary D. D. M., N. L. Ruff, and R. L. Dyck. 1994. "Development, critical period plasticity, and adult reorganizations of mammalian somatosensory systems." *Current Opinion in Neurobiol.*, 4: 535–544.

O'Leary, D. D. M., B. B. Stanfield, and W. M. Cowan. 1981. "Evidence that the early postnatal restriction of the callosal projection is due to the elimination of axon collaterals rather than to the death of neurons." *Dev. Brain Res.*, 1: 607–617.

Olivier, E., S. A. Edgley, J. Armand, and R. N. Lemon. 1997. "An electrophysiological study of the postnatal development of the corticospinal system in the Macaque monkey." *J. Neurosci.*, 17: 267–276.

Olmstead, C. E., J. R. Villablanca, B. J. Sonnier, J. P. McAlester, and F. Gomez. 1983. "Reorganization of cerebellorubral terminal fields following hemispherectomy in adult cats." *Brain Res.*, 274: 336–340.

Olson, C. R., and R. D. Freeman. 1980. "Cumulative effect of brief daily periods of monocular vision on kitten striate cortex." *Exp. Brain Res.* 38: 53–56.

O'Rahilly, R., and F. Müller. 1994. *The Embryonic Human Brain: An Atlas of Developmental Stages.* New York: Wiley-Liss.

Oski, F. A. 1993. "Iron deficiency in infancy and childhood." *N. Engl. J. Med.*, 329: 190–193.

Pantev, C., and B. Lutkenhoner. 2000. Magnetoencephalographic studies of functional organization and plasticity of the human auditory cortex. *J. Clin. Neurophysiol.*, 17: 130–142.

Pantev, C., R. Oostenveld, A. Engelien, B. Ross, L. E. Roberts, and M. Hoke. 1998. "Increased auditory representation in musicians". *Nature*, 392: 811–814.

Pascual-Leone, A., J. R. Gates, and A. Dhuna. 1991. "Induction of speech arrest and counting errors with rapid-rate transcranial magnetic stimulation." *Neurology*, 41: 697–702.

Pascual-Leone, A., C. M. Houser, K. Reese, L. I. Shotland, J. Grafman, S. Sato, J. Valls-Solé, J. P. Brasil-Neto, E. M. Wassermann, L. G. Cohen, and M. Hallett. 1993. "Safety of rapid-rate transcranial magnetic stimulation in normal volunteers." *Electroencephalogr. Clin. Neurophysiol.*, 89: 120–130.

Pasqual-Leone, A., F. Tarazona, J. Keenan, J. M. Tormos, R. Hamilton, and M. D. Catala. 1999. "Transcranial magnetic stimulation and neuroplasticity." *Neuropsychologia*, 37: 207–217.

Pascual-Leone, A., J. Valls-Solé, E. M. Wassermann, and M. Hallett. 1994. "Responses to rapid-rate transcranial magnetic stimulation of the human motor cortex." *Brain,* 117: 847–858.

Passingham, R. E., V. H. Perry, and F. Wilkinson. 1983. "The long-term effects of removal of sensorimotor cortex in infant and adult Rhesus monkeys." *Brain,* 106: 675–705.

Patterson, P. H. 1978. "Environmental determination of autonomic neurotransmitter functions." *Ann. Rev. Neurosci.,* 1: 1–17.

Paus, T., A. Zijdenbos, K. Worsley, D. L. Collins, J. Blumenthal, J. N. Giedd, J. L. Rapoport, and A. C. Evans. 1999. "Structural maturation of neural pathways in children and adolescents: in vivo study." *Science,* 283: 1908–1911.

Pearson, M. A., J. E. Hoyme, L. H. Seaver, and M. E. Rimsza. 1994. "Toluene embryopathy: delineation of the phenotype and comparison with fetal alcohol syndrome." *Pediatrics,* 93: 211–215.

Penfield, W., and T. Rasmussen. 1950. *The Cebral Cortex of Man.* New York: Macmillan.

Pennington, B. F. 1991. "Genetic and neurological influences on reading disability: an overview." *Reading and Writing,* 3: 191–201.

Perani, D., E. Paulesu, N. S. Galles, E. Dupoux, S. Dehaene, V. Bettinardi, S. F. Cappa, F. Facio, and J. Mehler. 1998. "The bilingual brain: proficiency and age of acquisition of the second language." *Brain,* 121: 1841–1852

Perlstein, M. A., and P. N. Hood. 1954. "Infantile spastic hemiplegia. III: intelligence." *Pediatrics,* 15: 676–682.

Petersen, S. E., P. Fox, M. Posner, M. Mintun, and M. Raichle. 1988. "Positron emission tomographic studies of the cortical anatomy of single-word processing." *Nature,* 331: 585–589.

Pfefferbaum, A., D. H. Mathalon, E. V. Sullivan, J. M. Rawles, R. B. Zipursky, and K. O. Lim. 1994. "A quantitative magnetic resonance imaging study of changes in brain morphology from infancy to late adulthood." *Arch. Neurol. Sep.,* 51 (9): 874–887.

Piccini, P., O. Lindvall, A. Bjorklund, P. Brundin, P. Hagell, R. Ceravolo, W. Oertal, N. Quinn, M. Samuel, S. Rehncrona, H. Widner, and D. J. Brooks. 2000. "Delayed recovery of movement-related cortical function in Parkinson's disease after striatal dopaminergic grafts." *Ann. Neurol.,* 48:689–695.

Pietro, S., V. L. Towle, R. Cakmur, and J. P. Spire. 1997. "Maturation of human visual evoked potentials: 27 weeks conceptional age to 2 years." *Neuropediatrics*, 28: 1–6.

Pinker, S. 1994. *The Language Instinct*. New York: Morrow.

Poeppel, D. 1996. "A critical review of PET studies of phonological processing." *Brain Lang.*, 55: 317–351, discussion on 352–385.

Poliakov, G. I. 1961. "Some results of research into the development of the neuronal structure of the cortical ends of the analyzers in man." *J. Comp. Neurol.*, 117: 197–212.

Pons, T. P., P. E. Garraghty, A. K. Ommaya, J. H. Kaas, E. Taub, and M. M. Mishkin. 1991. "Massive cortical reorganization after sensory deafferentation in adult Macaques." *Science*, 252: 1857–1860.

Provis, D. M., D. van Driel, F. A. Bilson, and P. Russell. 1985. "Human fetal optic nerve: overproduction and elimination of retinal axons during development." *J. Comp. Neurol.*, 238: 92–100.

Pugh, K. R., B. A. Shaywitz, S. E. Shaywitz, R. T. Constable, P. Skudlarski, R. K. Fulbright, R. A. Bronen, D. P. Shankweiler, L. Katz, J. M. Fletcher, and J. C. Gore. 1996. "Cerebral organization of component processes in reading." *Brain*, 119: 1221–1238.

Pujol, J., J. Deus, J. M. Losilla, and A. Capdevila. 1999. "Cerebral lateralization of language in normal left-handed people studied with functional MRI." *Neurology*, 52: 1038–1043.

Pujol, J., P. Vendrell, C. Junque, J. L. Marti-Vilalta, and A. Capdevila. 1993. "When does human brain development end? evidence of corpus callosum growth up to adulthood." *Ann. Neurol.*, 34: 71–75.

Purves, D., and J. W. Lichtman. 1980. "Elimination of synapses in the developing nervous system." *Science*, 210: 153–157.

Pysh, J. J., and G. M. Weiss. 1979. "Exercise during development induces an increase in Purkinje cell dendritic tree size." *Science*, 206: 230–232.

Quartz, S. R., and T. J. Sejnowski. 1997. "The neural basis of cognitive development: a constructivist manifesto." *Behav. Brain Sci.*, 20: 537–596.

Rajapakse, J. C., C. DeCarli, A. McLaughlin, J. N. Giedd, A. L. Krain, S. D. Hamburger, and J. L. Rapoport. 1996. "Cerebral magnetic resonance image segmentation using data fusion." *J. Comput. Assist. Tomogr.*, 20 (2): 205–218.

Rakic, P. 1971. "Guidance of neurons migrating to the fetal monkey neocortex." *Brain Res.*, 33: 471–476.

——— 1985. "Limits of neurogenesis in primates." *Science*, 227: 1054–1056.

——— 1988. "Specification of cerebral cortical areas." *Science*, 241 (4862): 170–176.

——— 1998. "Young neurons for old brains?" *Nat. Neurosci.*, 1: 645–647.

Rakic, P., and J. P. Bourgeois. 1993. "Changing of synaptic density in the primary visual cortex of the Rhesus monkey from fetal to adult stage." *J. Neurosci.*, 13: 2801–2820.

Rakic, P., J. P. Bourgeois, M. F. Eckenhoff, N. Zecevic, and P. S. Goldman-Rakic. 1986. "Concurrent overproduction of synapses in diverse regions of the primate cerebral cortex." *Science*, 232: 232–235.

Rakic, P., J. P. Bourgeois, and P. S. Goldman-Rakic. 1994. "Synaptic development of the cerebral cortex: implications for learning, memory, and mental illness." *Prog. Brain Res.*, 102: 227–243.

Rakic P., and R. P. Riley. 1983. "Regulation of axon number in primate optic nerve by prenatal binocular competition." *Nature*, 305: 135–137.

Ramachandran, V. S. 1993. Behavioral and magnetoencephalographic correlates of plasticity in the adult human brain." *Proc. Natl. Acad. Sci. USA*, 90: 10413–10420.

Ramey, C. T., and F. A. Campbell. 1984. "Preventive education for high-risk children: cognitive consequences of the Carolina Abecedarian Project." *Amer. J. of Mental Deficiency*, 88: 515–523.

Ramey, C. T., and S. L. Ramey. 1994. "Which children benefit most from early intervention?" *Pediatrics*, 94: 1064–1066.

——— 1998. "Prevention of intellectual disabilities: early interventions to improve cognitive development." *Preventive Medicine*, 27: 224–232.

Rami, A., A. J. Patel, and A. Rabie. 1986a. "Thyroid hormone and development of the rat hippocampus: cell acquisition in the dentate gyrus." *Neuroscience*, 19: 1207–1216.

———— 1986b. "Thyroid hormone and development of the rat hippocampus: morphological alterations in granule and pyramidal cells." *Neuroscience*, 19: 1217–1226.

Ramon y Cajal, S. 1937. *Recollections of My Life*. Trans. H. Craigie from the 3rd Spanish ed. (1923). Philadelphia: American Philosophical Society.

———— 1960. "Studies on Vertebrate Neurogenesis." Trans. L. Guth. Springfield, Ill.: Thomas.

Ramsey, N. F., B. S. Kirkby, P. Van Gelderen, K. F. Berman, J. H. Duyn, J. A. Frank, V. S. Mattay, J. D. Van Horn, G. Esposito, C. T. Moonen, and D. R. Weinberger. 1996. "Functional mapping of human sensorimotor cortex with 3D BOLD fMRI correlates highly with $H_2$ $^{15}O$ PET rCBF." *J. Cereb. Blood Flow Metab.*, 16: 755–764.

Rankin, J. M., D. M. Aram, and S. J. Horwitz. 1981. "Language ability in right and left hemiplegic children." *Brain Lang.*, 14: 292–306.

Rauschecker, J. P. 1995a. "Compensatory plasticity and sensory substitution in the cerebral cortex". *Trends in Neurosci.*, 18: 36–43.

———— 1995b. "Developmental plasticity and memory." *Behav. Brain Res.*, 66: 7–12.

Rauscher, F. H., G. L. Shaw, and K. N. Ky. 1993. "Music and spatial task performance." *Nature*, 365: 611.

———— 1995. "Listening to Mozart enhances spatial-temporal reasoning: towards a neurophysiological basis." *Neurosci. Lett.*, 185: 44–47.

Rauscher, F. H., G. L. Shaw, L. J. Levine, E. L. Wright, W. R. Dennis, and R. L. Newcomb. 1997. "Music training causes long-term enhancement of preschool children's spatial-temporal reasoning." *Neurol. Res.*, 19: 2–8.

Reinoso, B. S., and A. J. Castro. 1989. "A study of cortical remodeling: analysis of corticospinal plasticity using retrograde fluorescent tracers in rats." *Exp. Brain Res.*, 74: 387–394.

Riva, D., and L. Cazzaniga. 1986. "Late effects of unilateral brain lesions before and after the first year of life." *Neuropsychologia*, 24: 423–428.

Rockel, A. J., R. W. Hiorns, and T. P. S. Powell. 1980. "The basic uniformity in structure of the neocortex." *Brain*, 103: 221–244.

Roessler, E., and M. Muenke. 1998. "Holoprosencephaly: a paradigm for the complex genetics of brain development." *J. Inherited Metabolic Disease*, 21: 481–497.

Rothman, S. M., and J. W. Olney. 1986. "Glutamate and the pathophysiology of hypoxic-ischemic brain damage." *Ann. Neurol.*, 19: 105–111.

Rousselot, P., C. Lois, and A. Alvarez-Buylla. 1995. "Embryonic (PSA) N-CAM reveals chains of migrating neuroblasts between the lateral ventricle and the olfactory bulb of adult mice." *J. Comp. Neurol.*, 351 (1): 51–61.

Sadato, N., A. Pascual-Leone, J. Grafman, V. Ibanez, M. P. Deiber, G. Dold, and M. Hallet. 1996. "Activation of primary visual cortex by Braille reading in blind subjects." *Nature*, 380: 526–528.

Salamon, G. 1990. *Magnetic Resonance Imaging of the Human Brain: An Anatomical Atlas.* New York: Raven Press.

Sarnthein, J., A. von Stein, P. Rappelsberger, H. Petsche, F. H. Rauscher, and G. L. Shaw. 1997. "Persistent patterns of brain activity: an EEG coherence study of the positive effect of music on spatial-temporal reasoning." *Neurol. Res.*, 19: 107–116.

Sauer B., G. Kammradt, I. Krauthausen, H. T. Kretschmann, H. W. Lange, and F. Wingert. 1983. "Qualitative and quantitative development of the visual cortex in man." *J. Comp. Neurol.*, 214: 441–450.

Savage-Rumbaugh, E. S., J. Murphy, R. A. Sevcik, K. E. Brakke, S. L. Williams, and D. M. Rumbaugh. 1993. "Language comprehension in ape and child." *Monogr. Soc. Res. Child Dev.*, 58: 1–212.

Scarr, S. 1997. "Why child care has little impact on most children's development." *Current Directions in Psychological Science*, 6: 143–148.

Schade, J. P., H. Van Backer, and E. Colon. 1964. "Quantitative analysis of neuronal parameters in the maturing cerebral cortex." *Prog. Brain Res.*, 4: 150–175.

Schade, J. P., and D. B. van Groenigen. 1961. "Structural organization of the human cerebral cortex. I. Maturation of the middle frontal gyrus." *Acta Anat.*, 47: 74–111.

Schilder, A. G., J. G. Van Manen, G. A. Zielhuis, E. H. Grievink, S. A. Peters, P. Van Den Broek. 1993. "Long-term effects of otitis media with effusion on language, reading, and spelling." *Clin. Otolaryngol.*, 18 (3): 234–241.

Schlaggar, B. L., and D. D. O'Leary. 1991. "Potential of visual cortex to develop an array of functional units unique to somatosensory cortex." *Science,* 252: 1556–1560.

Schmanke, T. D., J. R. Villablanca, V. Lekht, and H. M. Patel. 1998. "A critical period for reduced brain vulnerability to developmental injury. II. Volumetric study of the neocortex and thalamus in cats." *Brain Res., Dev. Brain Res.,* 105: 325–337.

Schneider, G. E. 1979. "Is it really better to have your brain lesion early? A revision of the Kennard Principle." *Neuropsychologia,* 17: 557–583.

Schull, W. J. 1998. "The somatic effects of exposure to atomic radiation: the Japanese experience, 1947–1997." Proc. Natl. Acad. Sci. *USA,* 95: 5437–5441.

Schull, W. J., and M. Otake. 1999. "Cognitive function and prenatal exposure to ionizing radiation." *Teratology,* 59: 222–226.

Seitz, R. J., N. P. Azari, U. Knorr, F. Binkofski, H. Herzog, and H. J. Freund. 1999. "The role of diaschisis in stroke recovery". *Stroke,* 30:1844–1850.

Seitz, R. J., P. Hoflich, F. Binkofski, L. Tellmann, H. Herzog, and H. J. Freund. 1998. Role of the premotor cortex in recovery from middle cerebral artery infarction. *Arch. Neurol.,* 55: 1081–1085.

Selnes, O. A. 1999. "Recovery from aphasia: activating the 'right' hemisphere." *Ann. Neurol.,* 45: 419–420.

Sharp, F. R., and K. L. Evans. 1983. "Bilateral [$C^{14}$] 2-deoxyglucose uptake by motor pathways after unilateral neonatal cortex lesions in the rat." *Dev. Brain Res.,* 6: 1–11.

Shaywitz, B. A., S. E. Shaywitz, K. R. Pugh, R. T. Constable, P. Skudlarski, R. K. Fulbright, R. A. Bronen, J. M. Fletcher, D. P. Shankweiler, L. Katz, and J. S. Gore. 1995. "Sex differences in the functional organization of the brain for language." *Nature,* 373: 607–609.

Shaywitz, S. E., B. A. Shaywitz, J. M. Fletcher, and M. D. Escobar.1990. "Prevalence of reading disability in boys and girls: results of the Connecticut Longitudinal Study." *JAMA,* 264: 998–1002.

Shaywitz, S. E., B. A. Shaywitz, K. R. Pugh, R. K. Fulbright, P. Skudlarski, W. E. Mencl, R. T. Constable, F. Naftolin, S. F. Palter, K. E. Marchione, L. Katz, D. P. Shankweiler, J. M. Fletcher, C. Lacadie, M. Keltz, and J. C. Gore.

1999. "Effect of estrogen on brain activation patterns in postmenopausal women during working memory tasks." *JAMA*, 281: 1197–1202.

Sherwin, B. 1997. "Estrogen effects on cognition in menopausal women." *Neurology*, 48 (suppl. 7): S21–26.

Shevell, M. I. 1999. "Guest editor: new technologies in pediatric neurology." *Seminars in Pediatr. Neurol.*, 6: 67–127.

Sidman, R. L., and P. Rakic. 1973. "Neuronal migration with special reference to developing human brain." *Brain Res.*, 62: 1–35.

Silveri, M. C., M. G. Leggio, and M. Molinari. 1994. "The cerebellum contributes to linguistic production: a case of agrammatic speech following a right cerebellar lesion." *Neurology*, 44: 2047–2050.

Simons, D., and S. Finger. 1984. "Some factor-affecting behavior after brain damage early in life." In *Early Brain Damage*, vol. 2, ed. S. Finger and C. R. Almli, 327–347. New York: Academic Press.

Singer, W., and F. Tretter. 1976. "Receptive-field properties and neuronal connectivity in striate and parastriate cortex of contour-deprived cats. *J. Neurophysiol.*, 39: 613–630.

Sloper, J. J., P. Brodal, and T. P. S. Powell. 1983. "An anatomical study of the effects of unilateral removal of sensorimotor cortex in infant monkeys on the subcortical projections of the contralateral sensorimotor cortex." *Brain*, 106: 707–716.

Small, S. L. 1994. "Connectionist networks and language disorders." *J. Communication Disorders*, 27: 305–323.

Smith, A., and O. Sugar. 1975. "Development of above normal language and intelligence 21 years after left hemispherectomy." *Neurology*, 25: 813–818.

Smith, E. and J. Jonides. 1999. "Storage and executive processes in the frontal lobes." *Science*, 283: 1657–1661.

Smith, S. D., P. M. Kelley, and A. M. Brower. 1998. "Molecular approaches to the genetic analysis of specific reading disability." *Human Biol.*, 70: 239–256.

Solis, M. M., and A. J. Doupe. 1997. "Anterior forebrain neurons develop selectivity by an intermediate stage of birdsong learning." *J. Neurosc.i*, 17: 6447–6462.

———— 1999. "Contributions of tutor and bird's own song experience to neural selectivity in the songbird anterior forebrain." *J. Neurosci.*, 19: 4559–4584.

Solodkin, A., and G. W. Van Hoesen. 1996. "Entorhinal cortex modules of the human brain." *J. Comp. Neurol.*, 365 (4): 610–617.

Spinelli, D. N., F. E. Jensen, and G. Viana-di-Prisco. 1980. "Early experience effect on dendritic branching in normally reared kittens." *Exp. Neurol.*, 68: 1–11.

Spitz, R. A. 1945. "Hospitalism: an inquiry into the genesis of psychiatric conditions in early childhood." *Psychoanalytic Studies of the Child*, 1: 53–74.

Spitzka, E. A. 1907. "A study of the brains of six eminent scientists and scholars belonging to the American Anthropometric Society, together with a description of the skull of Professor E. D. Cope." *Trans. Am. Phil. Soc.*, 21: 175–308.

Stanfield, B. B., and D. D. M. O'Leary. 1985a. "Fetal occipital cortical neurons transplanted to rostral cortex develop and maintain a pyramidal tract axon." *Nature*, 313: 135–137.

———— 1985b. "The transient corticospinal projection from the occipital cortex during the postnatal development of the rat." *J. Comp. Neurol.*, 238: 236–248.

Stanfield, B. B., D. D. O'Leary, and C. Fricks. 1982. "Selective collateral elimination in early postnatal development restricts cortical distribution of rat pyramidal tract neurones. *Nature*, 298: 371–373.

Stein, Z., M. Susser, G. Saenger, and F. Marolla. 1972. "Nutrition and mental performance." *Science*, 178: 708–713.

———— 1975. *Famine and Human Development: The Dutch Hunger Winter of 1944–1945.* Oxford: Oxford University Press.

Sternberg, R. J., and J. S. Powell. 1983. "The development of intelligence." In *Handbook of Child Psychology*, 4th ed., ed. P. H. Mussen, 341–419. New York: J.Wiley.

Stiles, J., and R. Nass. 1991. "Spatial grouping activity in young children with congenital right and left hemisphere brain injury." *Brain and Cogn.*, 15: 201–222.

St. James-Roberts, I. 1981. "A reinterpretation of hemispherectomy data without functional plasticity of the brain." *Brain Lang.*, 13: 31–53.

Stryker, M. P., and W. A. Harris. 1986. "Binocular impulse blockade prevents the formation of ocular dominance columns in cat visual cortex." *J. Neurosci.,* 6: 2117–2133.

Stryker, M. P., and H. Sherk. 1975. "Modification of cortical orientation selectivity in the cat by restricted visual experience: a reexamination." *Science,* 190: 904–905.

Stuss, D. T., and D. F. Benson. 1986. *The Frontal Lobes.* New York: Raven Press.

Sulston, J. E., E. Schierenberg, J. G. White, and J. N. Thomson. 1983. "The embryonic cell lineage of the nematode *Caenorhabditis elegans.*" *Dev. Biol.,* 100: 64–119.

Taketa, K., Y. Nomura, H. Sakuma, T. Tagami, Y. Okuda, and T. Nakagawa. 1997. "MR assessment of normal brain development in neonates and infants: comparative study of T1 and diffusion-weighted images." *J. Comput. Assist. Tomogr.,* 21: 1–7.

Taub, E., and J. E. Crago. 1995. "Increasing behavioral plasticity following central nervous system damage in monkeys and man: a method with potential application to human developmental motor disability." In *Maturational Windows and Adult Cortical Plasticity,* ed. B. Julesz and I. Kovacs, 201–215. Reading, Mass.: Addison Wesley.

Taub, E., N. E. Miller, T. A. Novack, E. W. Cook III, W. D. Fleming, C. S. Nepomucino, J. S. Connell, and J. E. Crago. 1993. "Technique to improve chronic motor deficit after stroke." *Arch. Phys. Med. Rehab.,* 74: 347–354.

Taylor, D., J. Vaegan, A. Morris, J. E. Rogers, and J. Warland. 1979. "Amblyopia in bilateral infantile and juvenile cataract: relationship to timing of current treatment." *Trans. Ophthal. Soc. UK,* 99: 170–175.

Teele, D. W., J. O. Klein, and B. A. Rosner. 1984. "Otitis media with effusion during the first three years of life and development of speech and language." *Pediatrics,* 74: 282–287.

Tees, R. C., and J. F. Werker. 1984. "Perceptual flexibility: maintenance or recovery of the ability to discriminate non- native speech sounds. *Canad. J. Psychol.,* 38: 579–590.

Teller, D. Y. 1983. "Scotopic vision, color vision, and stereopsis in infants." *Current Eye Research,* 2: 199–210.

Teuber, H. L. 1974. "Functional recovery after lesions of the nervous system. II. Recovery of function after lesions of the central nervous system: history and prospects." *Neurosci. Res. Program Bull.*, 12: 197–211.

Thal, D., V. Marchman, J. Stiles, D. Aram, D. Trauner, R. Nass, and E. Bates. 1991. "Early lexical development in children with focal brain injury." *Brain Lang.*, 40: 491–527.

Tomblin, J. B., L. Spencer, S. Flock, R. Tyler, and B. Gantz. 1999. "A comparison of language achievement in children with cochlear implants and children using hearing aids." *J. Speech Lang. Hear. Res.*, 42: 497–509.

Turner, A. M., and W. T. Greenough. 1985. "Differential rearing effects on rat visual cortex synapses. I. Synaptic and neuronal density and synapses per neuron." *Brain Res.*, 329: 195–203.

Tyler, R. S., H. Fryauf-Bertschy, D. M. Kelsay, B. J. Gantz, G. P. Woodworth, and A. Parkinson. 1997. "Speech perception by prelingually deaf children using cochlear implants." *Otolaryngology Head and Neck Surgery*, 117 (3, pt. 1): 180–187.

Tyler, R. S., B. J. Gantz, G. G. Woodworth, H. Fryauf-Bertschy, and D. M. Kelsay. 1997. "Performance of 2- and 3-year-old children and prediction of 4-year from 1-year performance." *Amer. J. Otology*, 18: S157–159.

Ueki, K. 1966. "Hemispherectomy in the human with special reference to the preservation of function." *Prog. Brain Res.*, 21B: 285–338.

Uno, H., L. Lohmiller, C. Thieme, et al. 1990. "Brain damage induced by prenatal exposure to dexamethasone in fetal Rhesus/Macaques: I. Hippocampus." *Dev. Brain Res.*, 53: 157–167.

Uylings, H. B. 2000. "Development of the cerebral cortex in rodents and man." *Eur. J. Morphol.*, 38: 309–312.

Uylings, H. B., K. Kuypers, and W. A. Veltman. 1978. "Environmental influences on the neocortex in later life." *Prog. Brain Res.*, 48: 261–274.

Uylings, H. B., and J. G. Parnavelas. 1981. "Growth and plasticity of cortical dendrites." In *Cellular Analogues of Conditioning and Neuronal Plasticity*, ed. O. Fehrer and F. Joo, 57–64. London: Pergamon Press.

Uylings, H. B., and C. G. Van Eden. 1990. "Qualitative and quantitative comparison of the prefrontal cortex in rat and in primates, including humans." *Prog. Brain Res.*, 85: 31–62.

Van der Knaap, M. S., J. Valk, C. J. Bakker, M. Schooneveld, J. A. Faber, J. Willemse, and R. H. Gooskens. 1991. "Myelination as an expression of the functional maturity of the brain." *Dev. Med. Child Neurol.*, 33: 849–857.

Van der Loos, H., and J. Dörfli. 1978. "Does the skin tell the somatosensory cortex how to construct a map of the periphery?" *Neurosci. Lett.*, 7: 23–30.

Van der Loos, H., E. Welker, and J. Dörfl. 1986. "Selective breeding for variations in patterns of mystacial vibrissae of mice." *J. Heredity*, 77: 66–82.

Van Dongen, H. R., C. E. Catsman-Berrevoets, and M. A. Marijka van Mourik. 1994. "The syndrome of 'cerebellar' mutism and subsequent dysarthria." *Neurology*, 44: 2040–2046.

Van Hof, M. W., J. F. Hobbelen, and W. H. De Vos-Korthals. 1987. "Motor behavior and visual discrimination after neonatal and adult hemidecortication in the rabbit." *Behav. Brain Res.*, 25: 247–253.

Vargha-Khadem, F., E. B. Isaacs, H. Papaleloudi, C. E. Polkey, and J. Wilson. 1991. "Development of language in 6 hemispherectomized patients." *Brain*, 114: 473–495.

Vargha-Khadem, F., A. O'Gorman, and G. Watters. 1985. "Aphasia and handedness in relation to hemispheric side, age at injury and severity of cerebral lesion during childhood." *Brain*, 108: 677–696.

Vargha-Khadem, F., and C. E. Polkey. 1992. "A review of cognitive outcome after hemidecortication in humans." In *Recovery from Brain Damage: Advances in Experimental Medicine and Biology*, vol. 325, ed. F. D. Rose and D. A. Johnson, 137–151.

Veraart, C., A. G. De Volder, M. C. Wanet-Defalque, A. Bol, C. H. Michel, and A. M. Goffinet. 1990. "Glucose utilization in human visual cortex is abnormally elevated in blindness of early onset but decreased in blindness of late onset." *Brain Res.*, 510: 115–121.

Verhaart, W. J. C., and W. Kramer. 1952. "The uncrossed pyramidal tract." *Acta Psychiat. Neurol. Scand.*, 27: 181–200.

Viana-di Prisco, G., and D. N. Spinelli. 1981. "Observations on cortical plasticity using horseradish peroxidase." *Exp. Neurol.*, 74: 935–939.

Villablanca, J. R., J. W. Burgess, and F. Benedetti. 1986. "There is less thalamic degeneration in neonatal-lesioned than in adult-lesioned cats after cerebral hemispherectomy." *Brain Res.*, 368: 211–225.

Villablanca, J. R., J. W. Burgess, and C. E. Olmstead. 1986. "Recovery of function after neonatal or adult hemispherectomy in cats: I. Time course, movement, posture, and sensorimotor tests." *Behav. Brain Res.*, 19: 205–226.

Villablanca, J. R., P. Carlson-Kuhta, T. D. Schmanke, and D. A. Hovda. 1998. "A critical maturational period of reduced brain vulnerability to developmental injury. I. Behavioral studies in cats." *Brain Res., Dev. Brain Res.*, 105: 309–324.

Villablanca, J. R., and F. Gómez-Pinilla. 1987. "Novel crossed corticothalamic projections after neonatal cerebral hemispherectomy: a quantitative autoradiography study in cats." *Brain Res.*, 410: 219–231.

Villablanca, J. R., F. Gómez-Pinilla, B. J. Sonnier, and D. A. Hovda. 1988. "Bilateral pericruciate cortical innervation of the red nucleus in cats with adult or neonatal cerebral hemispherectomy." *Brain Res.*, 453: 17–31.

Villablanca, J. R., D. A. Hovda, G. F. Jackson, and R. Gayek. 1993. "Neurological and behavioral effects of a unilateral frontal cortical lesion in fetal kittens. I. Brain morphology, movement, posture, and sensorimotor tests." *Behav. Brain Res.*, 57: 63–77.

Villablanca, J. R., D. A. Hovda, G. F. Jackson, and C. Infante. 1993. "Neurological and behavioral effects of a unilateral frontal cortical lesion in fetal kittens: visual system tests, and proposing an 'optimal developmental period' for lesion effects." *Behav. Brain Res.*, 57: 79–92.

Voeller, K. K. 1986. "Right-hemisphere deficit syndrome in children." *Amer. J. Psychiat.*, 143: 1004–1009.

Volkmar, F. R., and W. T. Greenough. 1972. "Rearing complexity affects branching of dendrites in the visual cortex of the rat." *Science*, 176: 1445–1447.

VonNoorden, G. K., and M. L. J. Crawford. 1979. "The sensitive period." *Trans. Ophthal. Soc. UK*, 99: 442–446.

Wada, J. A., R. Clarke, and A. Hamm. 1975. "Cerebral hemispheric asymmetry in humans." *Arch. Neurol.*, 32: 239–246.

Wall, P. D. 1977. "The presence of ineffective synapses and the circumstances which unmask them." *Phil. Trans. Roy. Soc. Lond. B*, 278: 361–372.

Wallace, I. F., J. S. Gravel, C. M. McCarton, D. R. Stapells, R. S. Bernstein, and R. J. Ruben. 1988. "Otitis media, auditory sensitivity, and language outcomes at one year." *Laryngoscope*, 98: 64–70.

Wallis, D. E., E. Roessler, U. Hehr, L. Nanni, T. Wiltshire, A. Richieri-Costa, G. Gillessen-Kaesbach, E. H. Zackai, J. Rommens, and M. Muenke. 1999. "Mutations in the homodomain of the human SIX3 gene cause holoprosencephaly." *Nature Genetics*, 22: 196–198.

Walsh, C., and C. Cepko. 1988. "Clonally related cells show several migration patterns." *Science*, 241: 1342–1345.

Walsh, R. N. 1981. "Effects of environment complexity and deprivation on brain anatomy and histology: a review." *Int. J. Neurosci.*, 12: 33–51.

Wassermann, E. M. 1998. "Risk and safety of repetitive transcranial magnetic stimulation: report and suggested guidelines from the International Workshop on the Safety of Repetitive Transcranial Magnetic Stimulation, June 5–7, 1996." *Electroencephalogr. Clin. Neurophysiol.*, 108 (1): 1–16.

Weber-Fox, C. M., and H. J. Neville. 1996. "Maturational constraints on functional specializations for language processing: ERP and behavioral evidence in bilingual speakers." *J. Cogn. Neurosci.*, 8: 231–256.

Weiller, C., F. Chollet, K. J. Friston, R. J. Wise, and R. S. Frackowiak. 1992. "Functional reorganization of the brain in recovery from striatocapsular infarction in man." *Ann. Neurol.*, 31: 463–472.

Weintraub, S., and M. M. Mesulam. 1983. "Developmental learning disabilities of the right hemisphere: emotional, interpersonal, and cognitive components." *Arch. Neurol.*, 40: 463–468.

Welker, E., and H. VanderLoos. 1986. "Quantitative correlation between barrel field size and the sensory innervation of the whisker pad: a comparative study in six strains of mice bred for different patterns of mystacial vibrissae." *J. Neurosci.*, 6: 3355–3373.

Werker, J. F., and R. C. Tees. 1983. "Developmental changes across childhood in the perception of non-native speech sounds." *Canad. J. Psychol.*, 37: 278–286.

Wickett, J. C., P. A. Vernon, and D. H. Lee. 2000. "Relationship between factors of intelligence and brain volume." *Person. Individ. Diff.,* 29: 1095–1122.

Wiesel, T. N. 1982. "Postnatal development of the visual cortex and the influence of environment." *Nature,* 299: 583–591.

Wiesel, T. N., and D. H. Hubel. 1965. "Comparison of the effects of unilateral and bilateral eye closure on cortical unit responses in kittens." *J. Neurophysiol.,* 28: 1029–1040.

Wilson, H. R. 1988. "Development of spatiotemporal mechanisms in infant vision." *Vision Research,* 28: 611–628.

Winfield, D. A. 1981. "The postnatal development of synapses in the visual cortex of the cat and the effects of eyelid closure." *Brain Res.,* 206: 166–171.

Winick, M. 1970. "Nutrition, growth, and mental development: biological correlates." *Amer. J. Dis. Child.,* 120: 416–418.

Witelson, S. F., D. L. Kigar, and T. Harvey. 1999. "The exceptional brain of Albert Einstein." *The Lancet,* 353: 2149–2153.

Witelson, S. F,. and W. Pallie. 1973. "Left hemisphere specialization for language in the newborn: neuroanatomical evidence of asymmetry." *Brain,* 96: 641–647.

Wolff, P. H., I. Melngailis, M. Obregon, and M. Bedrosian. 1995. "Family patterns of developmental dyslexia. Part II: behavioral phenotypes." *Amer. J. Med. Genet.,* 60: 494–505.

Wolpar, S. M., and P. D. Barnes. 1992. *MRI in Pediatric Neurosurgery.* St. Louis: Mosby.

Woods, B. T., and S. Carey. 1979. "Language deficits after apparent clinical recovery from childhood aphasia." *Ann. Neurol.,* 6: 405–409.

Woods, B. T., and H. L. Teuber. 1978. "Changing patterns of childhood aphasia." *Ann. Neurol.,* 3: 273–280.

Yakovlev, P. I., and A. R. Lecours. 1967. "The myelogenetic cycles of regional maturation of the brain." In *Regional Development of the Brain in Early Life,* ed. A. Minkowsky, 3–70. Oxford and Edinburgh: Blackwell.

Zatorre, R. J., A. C. Evans, and E. Meyer. 1994. "Neural mechanisms underlying melodic perception and memory for pitch." *J. Neurosci.,* 14: 1908–1919.

Zatorre, R. J., D. W. Perry, C. A. Beckett, C. F. Westbury, and A. C. Evans. 1998. "Functional anatomy of musical processing in listeners with absolute pitch and relative pitch. *Proc. Natl. Acad. Sci. USA*, 95: 3172–3177.

Zatorre, R. J., and S. Samson. 1991. "Role of the right temporal neocortex in retention of pitch in short-term auditory memory." *Brain*, 114: 2403–2417.

Z'Graggen, W. J., G. A. Metz, G. L. Kartje, M. Thallmair, and M. E. Schwab. 1998. "Functional recovery and enhanced corticofugal plasticity after unilateral pyramidal tract lesion and blockade of myelin-associated neurite growth inhibitor in adult rats." *J. Neurosci.*, 18: 4744–4757.

# AUTHOR INDEX

# SUBJECT INDEX

corticosteroids, prenatal exposure to, 19

critical period, 20, 22, 29, 34, 67, 68, 91, 93, 94, 96, 97, 100, 106, 124, 152, 153, 154, 164, 176, 207, 208, 215

crossed cortico-spinal projections, 115

cross-modal plasticity, 81, 82, 99, 101, 107

crowding effects, 118, 144, 193, 204, 214

crowding hypothesis, 99

dendritic development, 24

dendritic spines, 26, 37, 60

dendritic sprouting, 173, 179, 180, 181; in aging, 28

dentate gyrus, 19, 22, 173, 181

developmental handicaps, 164

diffusion tensor imaging (DTI), 78

distributed system, 141, 162, 189, 191, 200

dopamine, 73, 74

EEG, 69, 78, 79

Einstein's brain, 145, 178, 179, 180, 182

elective brain functions, 120, 155, 156, 157, 197

enrichment programs, 83, 163, 164, 165, 177, 201, 207, 209, 215

environmental effects: on intelligence, 163; on language development, 145

environmental enrichment, 65, 177, 178, 197, 208

environmentalist position, 194

environmental stimulation, 16, 40, 72, 83, 85, 190, 203, 207, 208, 210

equipotentiality, 194, 195, 199, 204

event-related potential (ERP), 80, 81, 140, 199

evoked potentials, 80, 81, 95, 104, 136, 139

excitotoxicity, 128, 183

executive functions, 63, 155, 161, 162

experience dependent plasticity, 175, 176

experience expectant plasticity, 175, 176

face recognition, 200

facial movements, sparing of, 121

febrile seizures, 26

feral children, 156, 175

fetal alcohol syndrome, 19

fetal lesions: in felines, 168; in monkeys, 168

filamin 1 gene, 16

folic acid, 17, 20; supplementation, 17

functional magnetic resonance imaging (fMRI), 69, 72, 75–78, 80–83, 86, 108, 123, 126–129, 136, 138, 140, 142, 143, 144, 148, 149, 156, 158, 162, 180, 181, 199, 200, 202, 211, 213, 215

functional stabilization, 40

functional validation, 85, 196

fusiform gyrus, 200, 201

GABAnergic synapses, 65

gender-related differences in the cortical representation of language, 144

germinal matrix, 11, 12, 15, 173

glutamatergic synapses, 65, 74, 75

Golgi method, 24–27, 43, 173, 177, 180

hearing loss in infancy, 102

hemispherectomy, 112, 114, 116, 118, 119, 121, 122, 127, 128, 131, 134, 135, 136, 166, 167, 184, 186, 187, 188, 204

hierarchical synaptic development, 61

hippocampus, 19, 22, 35, 43, 61, 63, 152, 173, 181

holoprosencephaly, 15

homeobox genes, 15

homunculus, 104, 106, 129

hormonal effects: on brain development, 144; on cognitive functions, 158; on language development, 152

hypothyroidism, 34, 35, 64

idiot savant, 214

innatist position, 197, 199, 200, 201

intelligence, 163
iron deficiency, 19

language comprehension, 49, 132, 134, 150, 175
language development, 85, 103, 131, 134, 145, 146, 147, 152, 153, 154, 175, 199, 201; precocious, 140
language in the congenitally deaf, 148
language learning, 104, 110, 132, 139, 140, 141, 146, 149, 150, 151, 153, 164, 176, 186, 207, 208
language production, 66, 133, 134, 135
lesion-induced plasticity, 86, 111, 153, 186
lightning calculators, 214
limb restraint, 124, 185
LIS-1 gene, 15, 16
lissencephaly, 15
London taxi drivers, 181

magnetic resonance imaging (MRI), 68, 78
magnetic stimulation, 69, 73, 81, 82, 83, 86, 100, 101, 127
magnetoencephalography (MEG), 69, 81, 86, 159, 215
magnocellular system, 94
malnutrition: maternal-fetal, 18; protein calorie, 18
marginal layer, 12, 13
megalencephaly, 54
mental retardation, 15, 18, 19, 21, 22, 34, 54, 64
microcephaly, 18, 19, 21, 23
mirror movements, 122, 123
mismatching of synaptic connections, 34
MK 801, 74
modular system, 141
Mozart effect, 161
music learning, 141, 158
muteness, 133, 135
mutism, 137, 183

myelin-associated axonal growth inhibitors, 29
myelination, 62, 63, 78; of subcortical white matter, 62; delayed, 63

neural induction, 10
neural tube formation, 10
neurogenesis, 15, 21, 23, 37, 153, 173, 181, 198, 213; in adult brain, 22
neurons: birth of, 11; migration of, 12, 21
neurotransmitters, 32, 39, 65, 73, 74, 86, 128, 183
nonverbal learning disability, 137

obligatory brain functions, 120
ocular dominance columns, 30, 32, 89, 90, 91, 92, 93, 97, 100, 193
olfactory cortex, 63

parallel processing, 162
paraphasias, 132
paretic hand, neglect of, 124
parvocellular system, 94, 168
positron emission tomography (PET) 68–74, 77, 80, 83, 86, 99, 100, 128, 129, 136, 138, 140, 142, 143, 162, 166, 183, 199, 215
phantom limb sensations, 108, 110, 182, 183
phenylketonuria, 22, 64, 208
phosphotungstic acid method, 43
planum temporale, 145, 179
platelet activating factor, 13, 15
polymicrogyria, 21
prefrontal cortex, 12, 27, 28, 43–46, 48, 56, 61, 64, 67, 71, 72, 74, 75, 125, 126, 143, 156, 161, 162, 167, 168, 180, 186, 200, 202, 208, 209
pregnancy risk factors, 16
premature infants, 95, 164
pyramidal cell, 25, 33

radial glia, 12, 13
radiation exposure, 18, 21, 72, 77, 86
Rasmussen disease, 121, 137
Rett syndrome, 64

school effects, 84, 146, 153; on cognitive functions, 84; on language development, 84, 132, 146, 147
scientific writing, 139
second language learning, 149
serotonin, 73
signal to noise ratio, 141
silent synapses, 40, 41, 106, 110, 192
single unit recording, 89, 100
somatosensory system, 101, 104, 183, 188, 195
sound discrimination in infants, 150
squinting, 92, 93, 96, 97, 98, 99, 109, 110, 142, 208
strabismic amblyopia, 66, 96, 97, 109, 139, 182
strabismus, 95–99, 124, 142
stroke: adult, 183; pre- or perinatal, 121; gait in, 122
subplate neurons, 12, 13, 15, 35, 54
synaptic pruning, 26, 41, 44, 48, 50, 55, 59, 60, 71, 74, 80, 113, 190, 191, 192, 206, 207
synaptogenesis, 36, 37, 39, 40, 41, 43–45, 46, 49, 50, 55–62, 66, 71, 73, 74, 95, 116, 126, 157, 177, 198, 199, 209, 213

tabula rasa, 195
target finding, 29, 52, 53, 54
thymidine labeling, 12, 23
thyroid hormone, 21, 34, 36, 208
toluene, 19
transcranial magnetic stimulation (TMS), 81, 82, 100, 129
transient axons, 53, 115, 116, 117, 119, 122
trophic factors, 35, 39
tuberous sclerosis, 33

uncrossed cortico-spinal projections, 114
unmasking of synaptic connections, 108

visual deprivation, 58, 60, 89–94, 96, 98; bilateral, 90, 94; unilateral, 91, 96
vulnerable periods, 20

Wernicke aphasia, 132
Wernicke's area, 49, 56, 66, 104, 132–136, 139, 143, 145, 149, 150, 157, 183
West syndrome, 64
whiskers: removal of, 106; transplantation of, 10, 106, 110
windows of opportunity, 20, 111, 149, 207, 208
word spurt, 140